IGCSE
Mathematics Revision Guide

Martin Law

CAMBRIDGE
UNIVERSITY PRESS

CAMBRIDGE UNIVERSITY PRESS
Cambridge, New York, Melbourne, Madrid, Cape Town, Singapore, São Paulo

Cambridge University Press
The Edinburgh Building, Cambridge CB2 2RU, UK

www.cambridge.org
Information on this title: www.cambridge.org/9780521539029

First published 2004
4th printing 2006

Printed in the United Kingdom at the University Press, Cambridge

A catalogue record for this publication is available from the British Library

ISBN-13 978-0-521-53902-9 paperback
ISBN-10 0-521-53902-1 paperback

Past examination questions

Past examination questions are reproduced by permission of the University of Cambridge Local Examinations Syndicate: page 172,
practice extended exam 5, question 10, paper 4, syllabus code 0580/4, November 1996, question 7; page 167, exam 1, question 17,
paper 4, syllabus code 0580/4, November 1995, question 12; page 167, exam 2, question 1c, paper 2, syllabus code 0580/2,
November 1993, question 1; page 171, exam 5, question 3a, paper 2, syllabus code 0580/2, May 1989, question 6.

Contents

> * Topic 19 Straight line graphs becomes a **Core** topic for the first examination in 2006.
> ** The italicised areas of study are not examined until the first examination in 2006.

Curriculum outline

There are two curriculums in IGCSE Mathematics: **Core** and **Extended**. In this book the topics are covered in syllabus order, with the Core material explained first, followed if necessary by an additional section that covers the Extended material.

There are minor changes to the syllabus in topics 19, 33 and 34 which take effect from the first examination in 2006. Changes are noted in the text and additional material provided where necessary.

Core curriculum

The Core curriculum is for students aiming for grades C to G. At the end of the course the student will sit two separate exams, Papers 1 and 3:

- Paper 1 is one hour long. It consists of short-answer questions and is worth approximately 35% of the final grade.
- Paper 3 is two hours long. It consists of longer structured questions and is worth the remaining 65% of the final grade.

Extended curriculum

The Extended curriculum is for students aiming for grades A* to E. At the end of the course the student will sit two separate exams, Papers 2 and 4:

- Paper 2 is one and a half hours long. It consists of short-answer questions and is worth approximately 35% of the final grade.
- Paper 4 is two and a half hours long. It consists of longer structured questions and is worth the remaining 65% of the final grade.

All Extended curriculum students must cover the Core and Extended syllabus.

1 | Number, set notation and language

Use natural numbers, integers, prime numbers, common factors and multiples, rational and irrational numbers, real numbers, number sequences, recognise patterns in sequences, generalise to simple algebraic statements (i.e. *n*th term)

Natural numbers

The natural numbers are the counting numbers, i.e. 1, 2, 3, 4, etc.

Integers

Integers are whole numbers; they can be positive or negative, e.g. -5, 3, 25.

If the number contains a fraction part or a decimal point, then it cannot be an integer. (For example, the numbers 4.2 and $\frac{1}{2}$ are not integers.)

Prime numbers

Numbers that can only be divided by themselves and one, e.g. 2, 3, 5, 7, 11, 13, are prime numbers.

■ Note that 1 is not considered prime and 2 is the only even prime number.

Factors

A number is a factor of another number if it divides exactly into that number without leaving a remainder, e.g.

the factors of 6 are 1, 2, 3, 6
the factors of 15 are 1, 3, 5, 15

To find the factors of a number quickly, find which numbers were multiplied together to give that number.

For example, the products which give 8 are 1×8 or 2×4

so the factors of 8 are 1, 2, 4, 8

Highest Common Factor (HCF)

As the name suggests, this is simply the highest factor which is common to a group of numbers.

Example 1	Find the HCF of the numbers 6, 8 and 12.

Factors of 6 are   1, **2**, 3, 6
Factors of 8 are   1, **2**, 4, 8
Factors of 12 are 1, **2**, 3, 4, 6, 12
As the number 2 is a the highest factor of the three numbers, **HCF = 2**

Multiples

Multiples means 'the times table' of a number, e.g.   multiples of 4 are 4, 8, 12, 16, . . .
multiples of 9 are 9, 18, 27, 36, . . .

Lowest Common Multiple (LCM)

This is the lowest multiple which is common to a group of numbers. It is found by listing all the multiples of a group of numbers and finding the lowest number which is common to each set of multiples.

| Example 2 | Find the LCM of the numbers 2, 3, and 9. |

Multiples of 2 are 2, 4, 6, 8, 10, 12, 14, 16, **18**, 20 . . .
Multiples of 3 are 3, 6, 9, 12, 15, **18**, 21, 24 . . .
Multiples of 9 are 9, **18**, 27, 36 . . .

The number 18 is the lowest number that occurs as a multiple of each of the numbers. So the **LCM is 18**.

Rational numbers

Rational numbers are numbers which can be shown as fractions; they either terminate or have repeating digits, e.g. $\frac{3}{4}$, 4.333, 5.34 34 34, . . . etc.

■ Note that recurring decimals are rational.

Irrational numbers

An irrational number cannot be shown as a fraction, e.g. $\sqrt{2}$, $\sqrt{3}$, $\sqrt{5}$, π. Since these numbers never terminate, we cannot possibly show them as fractions.

The square root of any odd number apart from the square numbers is irrational. (Try them on your calculator, you will find that they do not terminate.) Also, any decimal number which neither repeats nor terminates is irrational.

For more information on square numbers see *Special number sequences* on page 4.

Number sequences

A number sequence is a set of numbers that follow a certain pattern, e.g.

1, 3, 5, 7, . . .   Here the pattern is either consecutive odd numbers or add 2.
1, 3, 9, 27, . . .   The pattern is 3 × previous number.

The pattern could be add, subtract, multiply or divide. To make it easier to find the pattern, remember that for a number to get bigger, generally you have to use the add or multiply operation. If the number gets smaller, then it will usually be the subtract or divide operation.

Sometimes the pattern uses more than one operation, e.g.

1, 3, 7, 15, 31, . . .   Here the pattern is multiply the previous number by 2 and then add 1.

The *n*th term

For certain number sequences it is necessary, and indeed very useful, to find a general formula for the number sequence.

Consider the number sequence 4, 7, 10, 13, 16, 19. We can see that the pattern is add 3 to the previous number, but what if we wanted the 50th number in the sequence?

This would mean continuing the sequence up to the 50th value, which is very time-consuming and indeed unnecessary.

A quicker method is to find a general formula for any value of *n* and then substitute 50 to find its corresponding value. These examples show the steps involved.

Example 3	Find the nth term and hence the 50th term of the following number sequence:

4   7   10   13   16   19   . . .

We can see that you add 3 to the previous number. To find a formula, follow the steps below.

Step 1   Construct a table and include a difference row.

n	1	2	3	4	5	6
Sequence	4	7	10	13	16	19

1st difference   3   3   3   3   3

Step 2   Look at the table to see where the differences remain constant.

We can see that the differences are always 3 in the first difference row. This means that the formula involves $3n$. If we then add 1, we get the original terms in the sequence:

When $n = 1$,   When $n = 2$
$3 \times (1) + 1 = 4$ $3 \times (2) + 1 = 7$

Step 3   Form a general nth term formula and check.

Knowing that we have to multiply n by 3 and then add 1:

nth term $= 3n + 1$

This formula is extremely powerful as we can now find the corresponding term in the sequence for any value of n. To find the 50th term in the sequence:

Using nth term $= 3n + 1$ when $n = 50$,
$3(50) + 1 = 151$

Therefore the 50th term in the sequence will be 151.

This is a much quicker method than extending the sequence up to $n = 50$.

Sometimes, however, we have sequences where the first difference row is not constant, so we have to continue the difference rows as shown in Example 4.

Example 4	Find the nth term and hence the 50th term for the following sequence:

0   3   8   15   24   35   . . .

Construct a table.

n	1	2	3	4	5	6
Sequence	0	3	8	15	24	35

1st difference   3   5   7   9   11

2nd difference   2   2   2   2

Now we notice that the differences are equal in the *second* row, so the formula involves n^2. If we square the first few terms of n we get: 1, 4, 9, 16, etc. We can see that we have to subtract 1 from these numbers to get the terms in the sequence. So

nth term $= n^2 - 1$

Now we have the nth term, to find the 50th term we use simple substitution:
$(50)^2 - 1 = 2499$

■ Note that some more complicated sequences will require a third difference row (n^3) for the differences to be constant, so we have to manipulate n^3 to get the final formula.

Special number sequences

Square numbers	1	4	9	16	25...	
	(1^2)	(2^2)	(3^2)	(4^2)	(5^2)	The counting numbers squared.

Cubed numbers	1	8	27	64	125...	
	(1^3)	(2^3)	(3^3)	(4^3)	(5^3)	The counting numbers cubed.

Triangular numbers   1   3   6   10   15...   Each number can be shown as a triangle, or simply add an extra number each time.

CORE   Number, set notation and language

1 | Number, set notation and language

Use language, notation and Venn diagrams to describe sets and represent relationships between sets

Definition of a set

A set is a collection of objects, numbers, ideas, etc. The different objects, numbers, ideas, etc. are called the **elements** or **members** of the set.

Example 1 Set A contains the even numbers from 1 to 10 inclusive. Write this as a set.

The elements of this set will be 2, 4, 6, 8, 10, so we write:

$A = \{2, 4, 6, 8, 10\}$

Example 2 Set B contains the prime numbers between 10 and 20 inclusive. Write this as a set.

The elements of this set will be 11, 13, 17 and 19. So

$B = \{11, 13, 17, 19\}$

$n(A)$, number of elements in set A

We count the number of elements in the set.

Example 3 If set C contains the odd numbers from 1 to 10 inclusive, find $n(C)$.

$C = \{1, 3, 5, 7, 9\}$

There are 5 elements in the set. So

$n(C) = 5$

$\in$, is an element of, and $\notin$, not an element of

The symbols $\in$ and $\notin$ indicate whether or not a certain number is an element of the set.

Example 4 Set $A = \{2, 5, 6, 9\}$. Describe which of the numbers 2, 3 or 4 are elements and which are not elements of set A.

Set A contains the element 2, therefore $2 \in A$.
Set A does not contain the elements 3 or 4, therefore $3, 4 \notin A$.

ξ, the universal set, and A′, the complement of a set

The universal set, ξ, for any problem is the set which contains all the available elements for that problem. The complement of a set is the set of elements of ξ which do not belong to A.

Example 5 The universal set is all of the odd numbers up to and including 11. List the universal set.

$\xi = \{1, 3, 5, 7, 9, 11\}$

Example 6 The complement of a set A is the set of elements of ξ in Example 5 which do not belong to A. List the complement of set A.

If $A = \{3, 5\}$ and $\xi = \{1, 3, 5, 7, 9, 11\}$,
then $A' = \{1, 7, 9, 11\}$

The empty set Ø or {}

The empty set contains no elements, e.g. for some pupils the set of people who wear glasses in their family will have no members.

■ Note that this is sometimes referred to as the null set.

Subsets A ⊆ B

If all the elements of a set A are also elements of a set B then A is said to be a **subset** of B. Every set has at least two subsets, itself and the null set.

Example 7	List all the subsets of {a, b, c}.

The subsets are Ø, {a}, {b}, {c}, {a, b}, {a, c}, {b, c} and {a, b, c}, because all of these elements can occur in their own right inside the main set.

■ Note that the number of subsets can be found by using the formula 2^n, where n = number of elements in the set.

Intersection A ∩ B and union A ∪ B

The **intersection** of two sets A and B is the set of elements which are common to both A and B. Intersection is denoted by the symbol ∩.

The **union** of the sets A and B is the set of all the elements contained in A and B. Union is denoted by the symbol ∪.

Example 8	If A = {2, 3, 5, 8, 9} and B = {1, 3, 4, 8} find: (a) A ∩ B (b) A ∪ B

(a) A ∩ B = {3, 8} These elements are common to both sets.

(b) A ∪ B = {1, 2, 3, 4, 5, 8, 9} These are the total elements contained in both A and B.

Venn diagrams

Set problems may be solved by using Venn diagrams. The universal set is represented by a rectangle and subsets of this set are represented by circles or loops. Some of the definitions explained earlier can be shown using these diagrams as follows:

A ∪ B

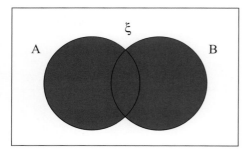

A ∩ B

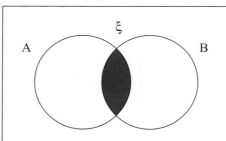

A′

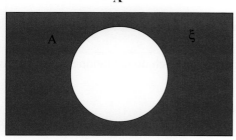

C ⊂ B

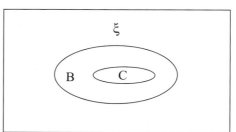

A ∩ B ∩ C

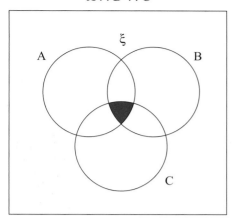

(A ∩ B) ∪ (A ∩ C)

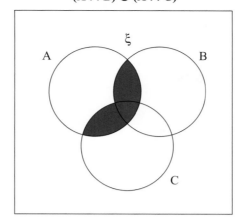

Obviously there are many different arrangements possible with these diagrams but now let us try some more difficult problems.

| Example 9 | $A = \{3, 4, 5, 6\}$, $B = \{2, 3, 5, 7, 9\}$ and $\xi = \{1, 2, 3, 4, 5, 6, 7, 8, 9, 10, 11\}$. Draw a Venn diagram to represent this information. Hence write down the elements of: |

(a) A' (b) $A \cap B$ (c) $A \cup B$

We only have two sets (A, B), so there are two circles inside the universal set:

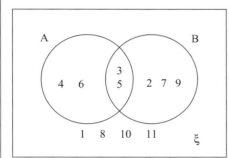

(a) $A' = \{1, 2, 7, 8, 9, 10, 11\}$

(b) $A \cap B = \{3, 5\}$

(c) $A \cup B = \{2, 3, 4, 5, 6, 7, 9\}$

Problems with the number of elements of a set

For two intersecting sets A and B we can use the rule:

$$n(A \cup B) = n(A) + n(B) - n(A \cap B)$$

This example shows how to use this rule.

| Example 10 | In a class of 25 members, 15 take history, 17 take geography and 3 take neither subject. How many class members take both subjects? |

Let H = set of History students, so $n(H) = 15$. Let G = set of Geography students, so $n(G) = 17$ and $n(H \cup G) = 22$ (since 3 students take neither subject). Let x represent the number taking both subjects.

Now we can draw the Venn diagram.

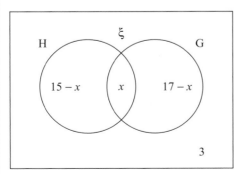

Using the formula:

$n(H \cup G) = n(H) + n(G) - n(H \cap G)$

$22 = 15 + 17 - x$

$x = 10$

Hence the number taking both subjects is 10.

The History only region is labelled $15 - x$. This is because a number of these students take Geography also.

■ Note that this formula can be used for any problem involving two sets.

2 | Squares, square roots and cubes

Know how to calculate squares, square roots and cubes of numbers

Squares (x^2)

The square of a number is that number multiplied by itself.

$$4^2 = 4 \times 4 = 16$$
$$8^2 = 8 \times 8 = 64$$

Square roots ($\sqrt{}$)

The inverse of a square can be quickly obtained from the calculator using the $\boxed{\sqrt{}}$ key.

$$\sqrt{16} = 4 \qquad \sqrt{49} = 7 \qquad \text{Since } 7 \times 7 = 49$$

Cubes (x^3)

The cube of a number is that number multiplied by itself twice.

$$5^3 = 5 \times 5 \times 5 = 125$$
$$10^3 = 10 \times 10 \times 10 = 1000$$

Cube roots ($\sqrt[3]{}$)

The inverse of a cube is obtained from the calculator using the $\boxed{\sqrt[3]{}}$ key.

$$\sqrt[3]{27} = 3 \qquad \text{Since } 3 \times 3 \times 3 = 27$$

Example 1 Find the value of: (a) $\sqrt{63} \times \sqrt{7}$ (b) $4^3 \div 2^3$

(a) $\sqrt{63} \times \sqrt{7} = 7.9372\ldots \times 2.6457\ldots$ The roots of 63 and 7 give continuing numbers after
$\phantom{\sqrt{63} \times \sqrt{7}} = 21$ the point so it's best to leave them in the calculator
display when multiplying them.

(b) $4^3 \div 2^3 \quad = 64 \div 8$ Four cubed and indeed two cubed can be done
$ = 8$ on the calculator, using the $\boxed{x^y}$ key or by simply
$ = 2^3$ multiplying the numbers together, i.e. $4 \times 4 \times 4 = 64$.

■ Note that when using the x^y key on your calculator, press the following:

$4^3 \div 2^3 = \boxed{4} \; \boxed{x^y} \; \boxed{3} \; \boxed{\div} \; \boxed{2} \; \boxed{x^y} \; \boxed{3} = 8$

For easy reference, these are the first 10 numbers and their corresponding squares and cubes:

x	1	2	3	4	5	6	7	8	9	10
x^2	1	4	9	16	25	36	49	64	81	100
x^3	1	8	27	64	125	216	343	512	729	1000

3 Directed numbers

Be able to apply directed numbers in practical situations (e.g. temperature change, flood levels)

Directed numbers are numbers with direction. Their sign, either positive or negative, indicates their direction. Positive numbers are above zero on the temperature scale or to the right of zero on a number line and negative numbers are below zero or to the left of zero on the number line, as illustrated in the diagrams below:

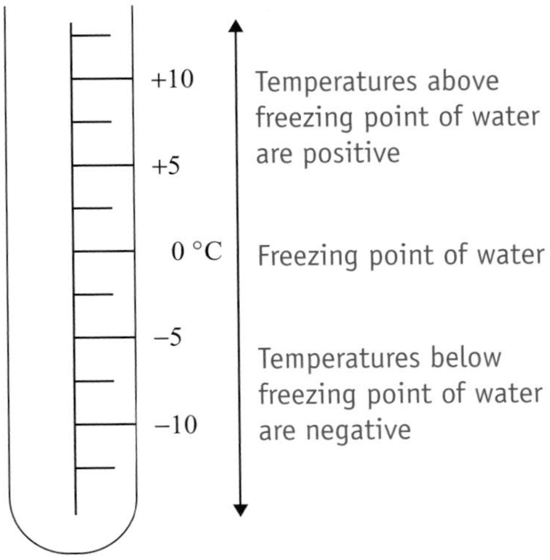

+10
+5
0 °C
−5
−10
Temperatures above freezing point of water are positive
Freezing point of water
Temperatures below freezing point of water are negative

A temperature scale

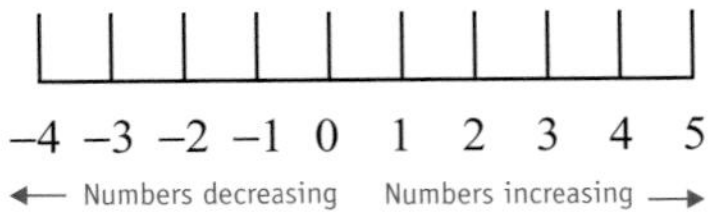

−4 −3 −2 −1 0 1 2 3 4 5
Numbers decreasing
Numbers increasing

A number line

A positive number displays either an upwards movement or a left-to-right movement. A negative number displays either a downward movement or a right-to-left movement.

Example 1 The diagram shows part of a water-level marker. The water level shown is -20 cm. What will be the reading on the marker if the water level: (a) rises by 30 cm (b) then falls by 17 cm?

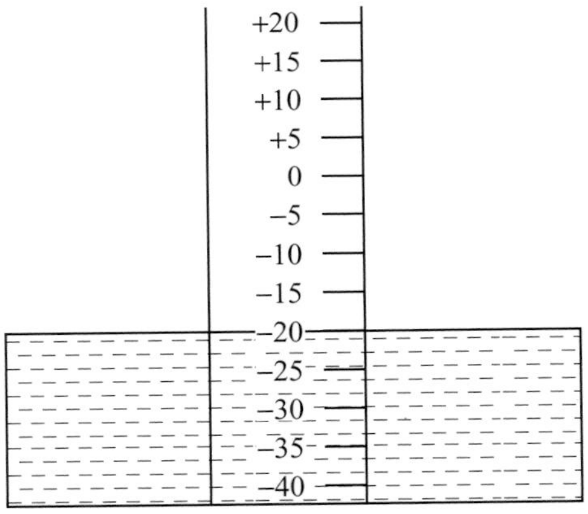

+20
+15
+10
+5
0
−5
−10
−15
−20
−25
−30
−35
−40

(a) If the water rises, this is a positive movement, so we add 30 to -20:

$-20\text{ cm} + 30\text{ cm} = 10\text{ cm}$

(b) The water then falls, so this is a negative movement and we have to subtract:

$10\text{ cm} - 17\text{ cm} = -7\text{ cm}$

Example 2 In July the average temperature in Saudi Arabia is 43 °C, while in Antarctica it averages -17 °C. Find the difference in temperature.

$43 - -17 = 60\ °\text{C}$

Note that we have to move 43 units down the scale to reach zero then another 17 degrees below zero to reach the point of -17 °C.

4 | Vulgar and decimal fractions and percentages

Understand and use simple vulgar and decimal fractions and percentages in appropriate contexts; recognise equivalence and convert between these forms

Fractions

A fraction is a part of something. The top number is called the **numerator** and the bottom number the **denominator**:

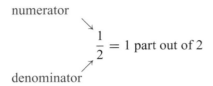

$$\frac{1}{2} = 1 \text{ part out of } 2$$

There are two other types of fraction:

- **Mixed fractions** contain a whole number and a fraction part.
- **Improper fractions** are 'top-heavy' fractions in which the numerator is bigger than the denominator.

Mixed and improper fractions can have the same value, but they are shown differently:

Mixed		Improper
$1\frac{2}{3}$	$\longrightarrow$	$\frac{5}{3}$
$2\frac{4}{7}$	$\longrightarrow$	$\frac{18}{7}$
$3\frac{3}{5}$	$\longrightarrow$	$\frac{18}{5}$
$4\frac{5}{6}$	$\longrightarrow$	$\frac{29}{6}$

Conversion between mixed and improper fractions

As an example, take $4\frac{2}{5}$. This can be changed to an improper fraction by multiplying the whole number by the fraction denominator and then adding the numerator part. This gives the new numerator of the top-heavy fraction. So

$$4\frac{2}{5} = \frac{(4 \times 5) + 2}{5} = \frac{22}{5}$$

For changing improper to mixed fractions, simply reverse the process:

$\frac{15}{4} = 15 \div 4 = 3$ wholes with 3 remainder, so the fraction is $3\frac{3}{4}$.

We can check our answers with a scientific calculator. Press these keys:

 $3\frac{3}{4}$

Fraction operations

This involves adding, subtracting, multiplying and dividing fractions.

Adding and subtracting fractions

If the denominators are the same, then just add or subtract the numerators.

Example 1   (a) $\frac{3}{7} + \frac{2}{7} = \frac{5}{7}$ (b) $\frac{5}{9} - \frac{2}{9} = \frac{3}{9} = \frac{1}{3}$ Simplify the final answer if possible.

If the denominators are different, then follow the steps shown in the example below.

Example 2   Find $\frac{2}{3} + \frac{3}{4}$

$\frac{8}{12} + \frac{9}{12} = \frac{17}{12}$

$= 1\frac{5}{12}$

Find the LCM (lowest common multiple) of the denominators. The LCM of 3 and 4 = 12.

Rewrite both of the fractions with 12 as the common denominator and add the numerators as before.

Simplify your answer.

The process is similar for subtracting, but when both the fractions have the same denominator, we *subtract* the numerators.

Multiplying fractions

First multiply the numerators together and then multiply the denominators together.

Example 3   Find $\frac{3}{8} \times \frac{4}{7}$

$\frac{3}{8} \times \frac{4}{7} = \frac{12}{56} = \frac{3}{14}$ Whenever possible, simplify the final answer.

Dividing fractions

First invert the second fraction and then multiply as normal.

Example 4   Find $\frac{2}{3} \div \frac{1}{12}$

$\frac{2}{3} \times \frac{12}{1} = \frac{24}{3}$

$= 8$

The second fraction has been inverted and then the numerators and denominators multiplied to give the result.

Finally simplify the answer.

Alternatively, we could 'cancel' the fractions: $\frac{2}{\cancel{3}} \times \frac{\cancel{12}^{4}}{1} = 2 \times 4 = 8$

■ Note that another name for inverting a fraction is finding its **reciprocal**.

The calculator method

The methods shown are useful to learn, but a great deal of time can be saved by using a scientific calculator.

First, locate the fraction key, usually the $\boxed{\text{a}^{\text{b/c}}}$ key and follow these steps:

$\frac{7}{9} \div \frac{2}{3} = \boxed{7}$ $\boxed{\text{a}^{\text{b/c}}}$ $\boxed{9}$ $\boxed{\div}$ $\boxed{2}$ $\boxed{\text{a}^{\text{b/c}}}$ $\boxed{3}$ $\boxed{=}$ $1\frac{1}{6}$

Decimal fractions and percentages

Fractions can be written as decimals, for example 0.5, 0.75. They can also be written as equivalent value fractions.

Percentages are fractions with denominators of 100. In solving many problems it is useful to express a fraction or a decimal as a percentage. Let's look at how to interchange fractions, decimals and percentages.

fraction ⟶ **decimal**	Divide numerator by denominator.

$\frac{3}{8} = 3 \div 8 = 0.375$

fraction or decimal ⟶ **percentage**	Multiply the fraction or decimal by 100.

(a) $\frac{4}{5} = \frac{4}{5} \times 100 = 80\%$ (b) $0.375 = 0.375 \times 100 = 37.5\%$

To convert back from percentage to decimal, simply reverse the process ($\div$ 100).

percentage ⟶ **fraction**	Place percentage value over 100, then simplify if possible.

(a) $30\% = \frac{30}{100} = \frac{3}{10}$ Divide by 10.   (b) $145\% = 1\frac{45}{100} = 1\frac{9}{20}$ Divide by 5.

Equivalent fractions

Equivalent fractions have the same value.

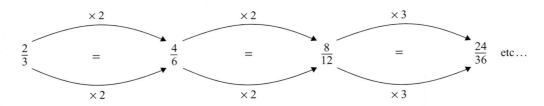

- Note that you must always do the same to the top (numerator) as to the bottom (denominator).

5 | Ordering

Order numbers using the symbols $=, \neq, >, <, \geq, \leq$

We need to know thoroughly the basic mathematical symbols that are used to compare sizes (magnitudes).

The symbols $=, \neq, >, <, \geq, \leq$

$a = b$ a equals b	$a \neq b$ a not equal to b
$a > b$ a is greater than b	$a < b$ a is less than b
$a \geq b$ a is greater than, or equal to, b	$a \leq b$ a is less than, or equal to, b

Now that we know the symbols, we can put different terms in order, as shown in these examples.

Example 1   Arrange the following terms in order of size, using the appropriate symbol:

$$0.55, \quad \tfrac{1}{2}, \quad 48\%$$

We have three terms shown in three different ways, i.e. a decimal, a fraction and a percentage. First we make them all decimals so that we can compare them more easily.

$0.55 = 0.55$ (already a decimal)
$\tfrac{1}{2} \quad = 1 \div 2 \quad = 0.5$
$48\% = 48 \div 100 = 0.48$

So $48\% < \tfrac{1}{2} < 0.55$.
     When we have all the numbers as decimals, we can see that $\tfrac{1}{2}$ is greater than 0.48 but smaller than 0.55, so we can put them in order.

Alternatively, we could have changed them all to percentages.

Example 2   Give all the possible values of x for the following expression:

$$3 \leq x < 8$$
     x must satisfy both sides of the expression, so x must be greater than or equal to 3 *and* x must be less than 8.

If x is less than 8, it could be 7, 6, 5, 4, 3, 2, etc.
If x is greater than or equal to 3, it could be 3, 4, 5, 6, 7, 8, 9, etc.
The values of x which satisfy both inequalities are 3, 4, 5, 6 and 7.   These numbers satisfy both sides of the expression.

---

6 | Standard form

Change numbers into standard form and vice versa

We use **standard form** (also called **scientific notation**) to show large or very small numbers in terms of powers of 10, for example we can write $10\,000\,000$ as 1×10^7 or $0.000\,000\,000\,001$ as 1×10^{-12}, which is a much more compact way to write them.

Rules for standard form

- Only one whole number between 1 and 9 inclusive is allowed before the decimal point.
- This number must be multiplied by a power of 10.

Changing large numbers into standard form

Look at the number $210\,000\,000$. We can change it into standard form:

$210\,000\,000$ is the same as $210\,000\,000.000$

As we are only allowed **one whole number** before the decimal point, we have to move the decimal point so that it appears immediately after the 2.

The decimal point has moved 8 places to the left.

The decimal point has moved 8 places to the left, so the new number is written as:

$$2.1 \times 10 \times 10 \times 10 \times 10 \times 10 \times 10 \times 10 \times 10 = 2.1 \times 10^8$$

Changing small numbers into standard form

Now look at the number $0.000\,003\,45$.

The decimal point has moved 6 places to the right.

The decimal point now move 6 places to the right, so as a power of ten it becomes:

$$3.45 \times 10^{-1} \times 10^{-1} \times 10^{-1} \times 10^{-1} \times 10^{-1} \times 10^{-1} = 3.45 \times 10^{-6}$$

Summary

- Numbers larger than 1 have a positive power of 10.
- Numbers smaller than 1 have a negative power of 10.
- Only one number, between 1 and 9, is allowed before the decimal point.
- Show all digits given in the question unless stated otherwise.

Example 1 Find the sum of 4×10^3 and 5×10^2, giving the answer in standard form.

First change the numbers to ordinary numbers: $4 \times 10^3 = 4000$
$5 \times 10^2 = 500$

Then we add them to get $4000 + 500 = 4500$.

Now change 4500 back into standard form: $4500 = 4.5 \times 10^3$ The decimal point is moved 3 places.

<table>
<tr><td>**Example 2**</td><td>Write the answers to the following calculations in standard form:</td></tr>
</table>

Example 2 Write the answers to the following calculations in standard form:
(a) $(2.5 \times 10^6) \times (3.0 \times 10^4)$ (b) $(4.5 \times 10^4) + (30.5 \times 10^3)$

We could use the same method as in Example 1, but they can also be calculated like this:

(a) $(2.5 \times 10^6) \times (3.0 \times 10^4) = 2.5 \times 3.0 \times 10^{6+4}$
$= 7.5 \times 10^{10}$

The numbers 2.5 and 3.0 are **multiplied** as normal to give 7.5.

The powers are **added** when multiplying terms (see Unit 23)

(b) $(4.5 \times 10^4) + (30.5 \times 10^3) = 4.5 \times 10^4 + (3.05 \times 10^4)$
$= 7.55 \times 10^4$

Make the powers the same by changing the second number.

Add the numbers as normal, taking the existing power.

The calculator method

Your scientific calculator can also be used to work out questions involving standard form. First find the key marked $\boxed{\text{EXP}}$, then follow the example below.

Example 3 Find the product of 2.34×10^2 and 4.56×10^{-1}.

Press the following keys on your calculator:

    $\boxed{2}$ $\boxed{\cdot}$ $\boxed{3}$ $\boxed{4}$ $\boxed{\text{EXP}}$ $\boxed{2}$ $\boxed{\times}$ $\boxed{4}$ $\boxed{\cdot}$ $\boxed{5}$ $\boxed{6}$ $\boxed{\text{EXP}}$ $\boxed{-1}$ $\boxed{=}$

You should get the answer: 106.704.

Standard form answers displayed on a calculator

When the result is either very large or very small (too many digits for the display), the calculator will give the answer in standard form. We need to be able to interpret the answer.

Result $= 5^{-03}$ means that 5 has been divided by 10 three times (3rd number after point) $= 0.005$.

Result $= 1.4336^{13}$ means 1.4336×10^{13}, an extremely large number!

If you find that results with relatively few digits are still being displayed in standard form, ask your teacher how to change the configuration of your calculator to amend this.

7 | The four rules

Calculate with whole numbers, decimals, BODMAS (correct ordering of operations)

The mathematical operations are add, subtract, multiply and divide. We shall only revise multiplication and division.

Long multiplication

It is important that we remember **place value** when carrying out long multiplication.
Take, for example, the number 234.68. Each of these digits has a certain value, as shown below.

	Hundreds	Tens	Units	Point	Tenths	Hundredths ...
	2	3	4	•	6	8
	↓	↓	↓		↓	↓
Value =	200	30	4	.	0.6	0.08

Example 1   Calculate 184×36.

$$
\begin{array}{r}
184 \\
\times\ 36 \\
\hline
1104 \\
+\ 5520 \\
\hline
6624 \\
\end{array}
$$

(184×6)
(184×30)
(184×36)

6 is in the units column, so we calculate $6 \times 184 = 1104$.
3 is in the tens column, so we put a zero in the units column and multiply by $3 = 30 \times 184 = 5520$.
Then we add them together to get 6624.

Short division

With short division, if there is a remainder, this is carried over to the next number.

Example 2   Calculate $451 \div 4$.

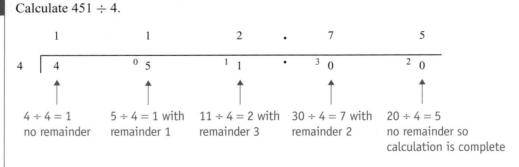

Answer = 112.75

■ Note that sometimes the answer is required in remainder form, i.e. 112 rem 3.

Long division

Long division is a more time-consuming method where each part of the calculation is shown under the previous part.

Example 3

Calculate $7165 \div 4$ to one decimal place.

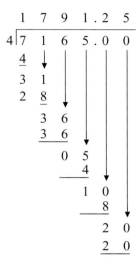

This time the remainders are shown after subtracting the multiples of 4. We then bring down the next digit (see arrows) to join the remainder and continue dividing by 4. Once we have no remainder and all of the original number has been divided by 4 we can stop the process.

Answer = 1791.3 (to 1 decimal place)

■ Note that some numbers will continue to give the same remainders (recurring decimals).

Operations with decimals

When adding or subtracting decimals, the most important rule is to make sure the numbers are in the right position relative to each other, i.e. the decimal points are in line under each other.

Example 4

Add 4.65 to 3.295.

$$
\begin{array}{r}
4.650 \\
+\ 3.295 \\
\hline
7.945
\end{array}
$$

The decimal points are lined up and hence the other numbers are also. The zero in the thousandths column for the first number is not strictly necessary for adding operations, but it is good practice to include it.

Example 5

Subtract 2.675 from 4.86.

$$
\begin{array}{r}
4.\cancel{8}7\ \cancel{6}^{1}5\ ^{1}0 \\
-\ 2.6\ \ 7\ \ \ 5 \\
\hline
2.1\ \ 8\ \ \ 5
\end{array}
$$

The highest number goes on top. Remember to add a zero.

We have borrowed and carried where required to get the final answer.

■ Note that if we are subtracting a decimal such as 4.36 from a whole number such as 25, we can rewrite 25 as 25.00. Now the place values line up and the calculation is straightforward.

Multiplying and dividing decimals

First multiply the whole numbers and then insert the decimal point as shown in the example. (It is good practice to make an estimate of the expected answer also.)

Example 6 | Calculate 0.65×2.1. | Estimate $= 1.3$.

```
          6    5
      × 2      1
 1    3    6    5
 1  . 3    6    5
```

The decimal point moves 3 places to the left.

First change each number to a whole number:

$$0.65 \longrightarrow 65 \quad \text{(decimal point moves 2 places)}$$
$$2.1 \longrightarrow 21 \quad \text{(decimal point moves 1 place)}$$

Perform the calculation using the whole numbers (21×65)

In total the point has moved $2 + 1 = 3$ places. When we get the answer, we simply move the point back 3 places to get the final answer. (1.365)

Example 7 | Calculate $4.26 \div 0.5$. | Estimate $= 8$.

```
            8 . 5  2
 50 | 4  2  6 . 0  0
```

Answer $= 8.52$

This time we have to move the decimal point 2 places in each number since the number with the greatest degree of accuracy is given to 2 decimal places:

$$4.26 \longrightarrow 426 \quad \text{(decimal point moves 2 places)}$$
$$0.5 \longrightarrow 50 \quad \text{(decimal point moves 2 places)}$$
$$\frac{4.26}{0.5} \longrightarrow \frac{426}{50}$$

Perform the calculation using the whole numbers.

■ Note that there is no need to move the decimal point in the answer with divide operations.

Correct ordering of operations (mixed calculations)

This is one of the most important topics you will learn. Every mixed calculation must follow certain steps in order to get the correct end answer. We use a method called **BODMAS** to tell us the importance of each operation in mathematics – that is, which operation to do first.

B	**O**	**D**	**M**	**A**	**S**
r	r	i	u	d	u
a	d	v	l	d	b
c	e	i	t		t
k	r	d	i		r
e		e	p		a
t			l		c
s			y		t

◀──────── More important   Less important ────────▶

The diagram above tells us that the most important operation is brackets, so we do this first if applicable, then deal with powers, followed by divide or multiply and then either add or subtract depending on the sign.

Example 8   Work out the value of:

(a) $19 \times 27 + 23 \times 8$ (b) $48 \div 6 - 54 \div 9$ (c) $11 - 12 \div 4 + 3(7 - 2)$

(a) $\underbrace{19 \times 27}_{\text{1st operation}} + \underbrace{23 \times 8}_{\text{1st operation}}$

As we have two multiply operations, these are performed at the same time, then the addition is carried out.

$= \underbrace{513 + 184}_{\text{2nd operation}}$

$= 697$

(b) $\underbrace{48 \div 6}_{\text{1st operation}} - \underbrace{54 \div 9}_{\text{1st operation}}$

We now have two divide operations. Perform these at the same time, then carry out the subtract operation.

$= \underbrace{8 - 6}_{\text{2nd operation}}$

$= 2$

(c) $11 - 12 \div 4 + 3\underbrace{(7 - 2)}_{\text{1st operation}}$

$= 11 - \underbrace{12 \div 4}_{\text{2nd operation}} + \underbrace{3 \times 5}_{\text{2nd operation}}$

There are brackets present so this must be the first operation, followed by the multiply and divide. Then we perform the addition and subtraction operations.

$= \underbrace{11 - 3}_{\text{3rd operation}} + 15$

$= \underbrace{8 + 15}_{\text{4th operation}}$

$= 23$

■ Note that the sign of the number is always to the left of it.

Any mixed calculation containing decimals or fractions can be performed using the same method.
Follow the rules of BODMAS.

8 | Estimation

Estimate numbers using rounding, decimal places and significant figures, and round off answers to a reasonable degree of accuracy for a given problem

Making estimates of numbers, quantities and lengths

In many cases, exact numbers are not necessary or even desirable. In these circumstances approximations are given instead. The most common types of approximations are:

- rounding
- decimal places (d.p.)
- significant figures (s.f.)

Rounding

This type of approximation is usually done for large numbers where the number itself is rounded to the required multiple of ten, i.e. nearest 10, nearest 100, nearest 1000, etc. Occasionally it can be used for small numbers, i.e. nearest whole number, nearest $\frac{1}{10}$, nearest $\frac{1}{100}$, etc.

Example 1 shows how a large number is rounded. Example 2 demonstrates a small number being rounded.

Example 1	In a recent football match there were 42 365 people attending. Approximate this number to: (a) nearest 10 (b) nearest 100 (c) nearest 1000.

(a) 42 365 to nearest 10 = 42 370	The number in the tens column is the 6. As the number after it is 5 or greater we round the 6 up to a 7.
(b) 42 365 to nearest 100 = 42 400	The number in the hundreds column is the 3. As the number directly after it is greater than 5 we round the 3 up to a 4.
(c) 42 365 to nearest 1000 = 42 000	The number in the thousands column is the 2. The number directly after it is less than 5, so we leave the 2 unchanged and simply replace the numbers following with zeros.

Example 2	Round the number 24.923 to: (a) nearest whole number (b) nearest tenth.

(a) 24.923 to nearest whole number = 25	The whole number is the unit or ones column, which in this case is the 4. As the number after it is greater than or equal to 5 we round up the 4 to a 5.
(b) 24.923 to nearest tenth = 24.9	The number 9 is in the tenths column and there is no need to round up.

Decimal places

This refers to the numbers after the decimal point. Very often a calculator gives us an answer that has too many decimal places. We can correct the answer to a given number of decimal places.

Example 3	Round the number 24.456 37 to (a) 3 decimal places (b) 2 d.p. (c) 1 d.p.

(a) 24.479 32 = 24.479 correct to 3 d.p.
     ↑↑↑
     123

We count to the third number after the decimal point, which in this case is the 9. Check if the number after it is 5 or greater. It is not, so the 9 stays unchanged. The rest of the digits are discarded.

(b) 24.479 32 = 24.48 correct to 2 d.p.
     ↑↑
     12

This time we count to the second number after the point. We have to round up the 7 to a 8.

(c) 24.479 32 = 24.5 correct to 1 d.p.
     ↑
     1

The first number after the point is the 4. This has to be rounded up to a 5.

Significant figures

Consider the number 24.67. The digit 2 is the most significant as it has the highest value, 20. The 4 is the next most significant, then the 6 and finally the 7, which is the least significant. We can therefore say that the number is given to **four significant figures**.

Some of the answers we obtain to problems contain too many figures for our purposes, so we need to approximate them to a given degree of accuracy just as we did for decimal places. This time the decimal point does not affect our final answer.

Example 4	Write 23.456 to: (a) 3 s.f. (b) 2 s.f. (c) 1 s.f.

(a) 2 3 . 4 5 6 = 23.5 correct to 3 s.f.
    ↑ ↑ ↑ ↑ ↑
    1 2 3 4 5
  significant figures

The third most significant figure is the 4. As the number after it is 5 or greater, we round it up to a 5.

(b) 2 3 . 4 5 6 = 23 correct to 2 s.f.

The second most significant figure is the 3. As the number after it is less than 5, there is no need to round this up.

(c) 2 3 . 4 5 6 = 20 correct to 1 s.f.

The 2 is the first significant figure. There is no need to round it up, but we must ensure the numbers are approximately the same size, so we need to add the zero after the 2 to make it 20.

As significant figures are more difficult, we will try another example.

Example 5	Write (a) 0.0046 and (b) 9.999 to 1 significant figure.

(a) 0.0046 = 0.005 (to 1 s.f.)
      ↑↑
      12

The first significant figure is the first digit which is not a zero, i.e. 4. The 4 then rounds up to a 5 as the number after it is 5 or greater.

(b) 9.999 = 10 (to 1 s.f.)

The first significant figure is the 9 which has to be rounded to the next highest figure which is a 10.

Giving answers to reasonable degrees of accuracy

On the front instruction page of all IGCSE papers, you are told by the examiner to give answers involving degrees to one decimal place and all other answers to three significant figures **unless stated otherwise**. If you come across a problem which does not state the degree of accuracy required, then follow a simple rule:

Give your answer to the same degree of accuracy as the question.

Example 6	Calculate $23.4 \div 4.56$, giving your answer to an appropriate degree of accuracy.
	$23.4 \div 4.56 = 5.13$ (to 2 d.p.) The greatest degree of accuracy in the question is 2 decimal places, so the answer must also be to 2 decimal places.

Example 7	A cuboid's dimensions are given as 3.873 metres by 2.5 metres by 3.2 metres. Calculate its volume, giving your answer to an appropriate degree of accuracy.
	Volume of cuboid = length $\times$ width $\times$ height = $3.873 \times 2.5 \times 3.2$ Three d.p. is as accurate $= 30.984 \, \text{m}^3$ as the question allows.

Example 8	Find the area of a right-angled triangle with base 4.3 cm and perpendicular height of 7 cm.
	Area $= \frac{1}{2}$ base $\times$ height $= \frac{1}{2} \times 4.3 \times 7 = 15.05 = 15.1 \, \text{cm}^2$ (to 1 d.p.) One d.p. is as accurate as the question allows.

Estimating answers to calculations

Many calculations can be done quickly using a calculator, but it is very easy to make mistakes when typing in the digits, so it useful to be able to make a quick check.

For example, estimate the answer to 58×207 to check the answer obtained using your calculator.

58 can be rounded to 60 and 207 can be rounded to 200, so now we have:

$60 \times 200 = 12\,000$ This is very close to our calculator answer of 12 006, so it is unlikely that we made a mistake on our calculator.

Summary

- When rounding answers to a given degree of accuracy, count up to the required number of places and then look at the number immediately after it to see if you need to round up. If the number immediately after is 5 or greater, round up.
- Unless stated otherwise, always give your answer to the same degree of accuracy as the question.
- Always make an approximate check on your calculator answer.
- When dealing with significant figures, remember that the approximated answer should be very similar in size to the original number. Add zeros if required to make the number of digits before the decimal point the same.
- With numbers less than 1, the first significant figure is the first digit which is not a zero.

9 | Limits of accuracy

Give upper and lower bounds for data given to a specified accuracy

All numbers can be written to different degrees of accuracy. For example, the number 5 is the same as 5.0 or 5.00, except that it has been shown to a different degree of accuracy each time.

If, for example, the number 3.5 has been rounded to one decimal place, then, before rounding, the number could have been anywhere in the range of 3.45 and up to but not including 3.55.

We could write this as an inequality:

$$3.45 \leq x < 3.55$$
 ↑ Lower bound   ↑ Upper bound

The lower bound is 3.45 and the upper bound is 3.55.

or even show it on a number line thus:

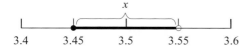

Note that the lower bound of 3.45 could be an actual value of x so it is drawn as a black circle, but the upper bound cannot be an actual value (as it would have to be rounded up to 3.6) so it is drawn as a clear circle.

Many students can find the upper and lower bounds quite easily in their head, but there is a method you can use if you find this difficult.

Example 1   A boy's height is given as 145 cm to the nearest centimetre. Find the upper and lower bounds within which his height could lie, giving your answer as an inequality.

The range is 1 cm, so if we first divide this by 2:   1 cm ÷ 2 = 0.5 cm.

Upper bound = 145 cm + 0.5 cm = 145.5 cm
Lower bound = 145 cm − 0.5 cm = 144.5 cm

The inequality is therefore 144.5 ≤ height < 145.5.

Example 2   The length of a field is measured as 90 metres to the nearest ten metres. Give the upper and lower bounds of the length.

10 m ÷ 2 = 5 metres
Upper bound = 90 + 5 = 95 metres
Lower bound = 90 − 5 = 85 metres

As an inequality: 85 ≤ length < 95.

---

9 | Limits of accuracy

Give appropriate upper and lower bounds to solutions of simple problems (e.g. the calculation of the perimeter or area of a rectangle) given data to a specified accuracy

If the numbers in a problem are written to a specified degree of accuracy, then the solutions to the problems involving those numbers will also give a range of possible answers.

Example 1 The circle below has a radius of 8 cm, measured to the nearest centimetre. Find the upper and lower bounds of its circumference.

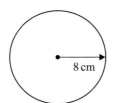

8 cm

Circumference of circle $= 2\pi r$
The radius lies in the range:

$$7.5\,\text{cm} \le r < 8.5\,\text{cm}$$

The circumference will also be between two ranges:

Upper bound of circumference $= 2 \times \pi \times 8.5 = 53.4\,\text{cm}$
Lower bound of circumference $= 2 \times \pi \times 7.5 = 47.1\,\text{cm}$

Example 2 Calculate the upper and lower bounds for the area of the rectangle shown if its dimensions are accurate to 1 decimal place.

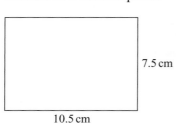

7.5 cm

10.5 cm

Area of rectangle $=$ length $\times$ width

Range of length : $10.45 \le l < 10.55$
Range of width : $7.45 \le w < 7.55$

Upper bound of area is the maximum value the area could be:

$$10.55 \times 7.55 = 79.65\,\text{cm}^2$$

Lower bound of area is the minimum value the area could be:

$$10.45 \times 7.45 = 77.85\,\text{cm}^2$$

Example 3 Calculate the lower and upper bounds for the number $\frac{54\,000}{600}$ if each of the numbers is given to 2 significant figures.

Range of 54 000 : $53\,500 \le 54\,000 < 54\,500$
Range of 600 : $595 \le 600 < 605$
Upper bound $= 54\,500 \div 595 = 91.60$
Lower bound $= 53\,500 \div 605 = 88.43$ (to 2 d.p.)

---

10 | Ratio, proportion, rate

Understand and use ratio, direct and inverse proportion and common measures of rate, use scales in practical situations, calculate average speed

Ratio

A ratio is used when we want to compare two or more quantities. For example, if $400 is to be shared between three people, Jim, Fred and John, in the ratio 4 : 3 : 1 respectively, the total amount of money is split into 8 equal parts ($4 + 3 + 1$), of which Jim will get 4 parts, Fred 3 parts and John 1 part as shown below:

$200 : $150 : $50   We can check all of the money has been accounted for by adding
 ↑   ↑   ↑   each of the amounts: $200 + $150 + $50 = $400 (original
Jim : Fred : John   amount).

Simplifying ratios

Many ratios are the same but shown differently, for example 1 : 2 is the same as 2 : 4. The second amount is twice the size of the first amount in each case.

In order to simplify ratios, we have to find common divisors of each part of the ratio:

          3 : 9   The ratio 3 : 9 simplifies to 1 : 3 since both three and nine divide by
÷ 3 ↓   ↓ ÷ 3   three exactly.
          1 : 3

■ Note that the ratios must be in the same units before you simplify them to their lowest terms. For example, to simplify the ratio 10 cm : 40 mm, we first make the units the same so it becomes 100 mm : 40 mm, then we divide both terms by 20 to get the answer 5 : 2.

Calculating ratios

There are two main types of ratio calculations:

- direct proportion
- inverse proportion

With **direct proportion**, as the share increases, then the amount also increases.

The example at the start of the page showed how Jim's amount of money was twice the value of Fred's amount because Jim's share was twice the value of Fred's share.

With **inverse proportion**, as the share increases, the actual amount decreases.

For example, if 8 people can dig a hole in 4 hours, how long would it take 2 people to dig the same hole? We now have the number of people reduced by a factor of 4 so therefore the time will increase by a factor of 4 to become 16 hours.

These examples involve direct proportion.

Example 1 | Share $200 in the ratio 3 : 2.

When calculating with ratios we can use either of the two methods shown:

<table>
<tr><td align="center">**1st method**</td><td align="center">**2nd method**</td></tr>
<tr><td>The first amount is 3 parts out of a total of 5 parts of the $200 and the second amount is 2 parts out of 5 parts of the $200:</td><td>Find the value of one part of the total by dividing the total money to be shared by the total number of parts, and then multiply each ratio by this singular amount:</td></tr>
</table>

$\frac{3}{5} \times \$200 = \$120$

$\frac{2}{5} \times \$200 = \$80$

One part $= \$200 \div 5 = \$40$

$3 \times \$40 = \$120$

$2 \times \$40 = \$80$

Answer from both methods is $120 : $80.

It is up to you which method you prefer, but either method can be extended to include more than two shares as shown in the following example.

Example 2 | $400 is distributed between four people in the ratios 2 : 3 : 4 : 7. Find the amount of money each person gets.

Using the first method: $\frac{2}{16} \times \$400 = \$50$ 2 parts of the total

$\frac{3}{16} \times \$400 = \$75$ 3 parts of the total

$\frac{4}{16} \times \$400 = \$100$ 4 parts of the total

$\frac{7}{16} \times \$400 = \$175$ 7 parts of the total

Total of parts $= 2 + 3 + 4 + 7$
$= 16$

Now make a simple check: $50 + $75 + $100 + $175 = $400   Correct total

Example 3 | An aircraft is cruising at 720 km/h and covers 1000 km. How far would it travel in the same period of time if the speed increased to 800 km/h?

We know that if an aircraft increases its speed when the time has remained constant, then its distance will also increase.

By comparing ratios:

Speed   720   :   800

Distance   1000 : x

$\dfrac{x}{800} = \dfrac{1000}{720}$

$x = \dfrac{800 \times 1000}{720} = 1111.1 \text{ km}$

In another type of ratio question the value of one of the amounts is given, and from this information you have to work out the value of the other amounts.

Example 4 | A sum of money is split between two people in the ratio of 2 : 3. If the smaller amount is $100, what is the value of the larger amount and the total sum of money altogether?

If we know that two parts of the total $= \$100$, then we can calculate the value of one part:

one part $= \$100 \div 2$ $= \$50$

so three parts $= 3 \times \$50$ $= \$150$

Total sum of money $= \$100 + \$150 = \$250$

■ Note how the second method has been used for this type of problem.

These examples involve inverse proportion.

Example 5 If 8 people can pick apples from some trees in 6 hours, how long would it take 12 people?

We know that the time taken to pick the apples will be less for 12 people than for 8 people, and that the time will decrease by the factor $\frac{12}{8}$.

Time taken $= 6 \div \frac{12}{8} = 4$ hours.

Example 6 A tap issuing water at a rate of 1.2 litres per minute fills a container in 4 minutes. How long would it take to fill the same container if the rate was decreased to 1 litre per minute?

If the flow is decreased by a factor of $\frac{1}{1.2}$ then the time should increase by the same factor.

Time $= 4 \div \frac{1}{1.2} = 4.8$ minutes.

Using scales in practical situations

Scales are used primarily in maps and model work. A typical map scale is given below:

1 : **25 000**   In this case, one unit on the scaled or map length is equal to 25 000 units in actual
↑   ↑   or real length.
Scaled   Actual
length   length

Example 7 A model car is $\frac{1}{40}$ scale model. Express this as a ratio.

If the length of the real car is 5.5 metres, what is the length of the model car? (Give your answer in centimetres.)

Ratio $= 1 : 40$ One unit on the model length is equal to 40 units on the actual car length.

Length of model car $= 5.5 \times \frac{1}{40} = 0.1375$ metres $= 13.75$ cm

Example 8 The scale of a map is $1 : 40\,000$.

Two villages are 8 cm apart on the map. How far apart are they in real life?

Real-life distance $= 8$ cm $\times 40\,000 = 320\,000$ cm $= 3.2$ km

Calculating average speed

Speed can be displayed in several ways: metres/second, km/h, cm/second, km/second. The units used depend on how fast or slow an object travels. For example, something which travels very quickly such as the speed of light would be displayed as 300 000 km/second, whereas the speed of a slow-moving object such as a snail would be represented in mm/second. Below are some useful conversion examples.

Example 9 A car is travelling at 25 metres/second. What speed does this represent in km/hour?

25 metres/second $= 0.025$ km/second   Divide by 1000 to change metres into km.
$= 90$ km/hour   Multiply by 60 and by 60 again to change seconds to hours.

Example 10 The speed of sound is 1188 km/h. Represent this in: (a) metres/second   (b) mm/minute

(a) 1188 km/h $= 1\,188\,000$ m/h   Multiply by 1000 to change km into metres. Divide by
$= 330$ m/second   60 and by 60 again to change hours into seconds.

(b) 1188 km/h $= 1\,188\,000\,000$ mm/h   Multiply by 1000 and by 1000 again to change km
into mm.

$= 19\,800\,000$ mm/minute   Divide by 60 to change hours into minutes.

In order to calculate speed, we need to know the distance and the time. We can use the following formula triangle to help us:

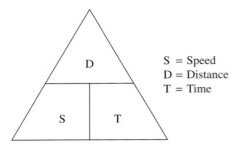

S = Speed
D = Distance
T = Time

$$\text{Speed} = \frac{\text{Distance}}{\text{Time}}$$
$$\text{Distance} = \text{Speed} \times \text{Time}$$
$$\text{Time} = \frac{\text{Distance}}{\text{Speed}}$$

It is useful to know all three formula. When performing calculations with speed, distance and time, always check that the units are the same.

Example 11 | If a cyclist travels 25 km in a time of 5 hours, find the average speed.

Speed = Distance ÷ Time = 25 ÷ 5 = 5 km/hour

Example 12 | Find the time taken for a car travelling at 120 km/hour to cover 360 km.

Time = Distance ÷ Speed = 360 ÷ 120 = 3 hours

Example 13 | What distance would a bus cover if it travelled for 4 hours at an average speed of 100 km/hour?

Distance = Speed × Time = 100 × 4 = 400 km

Example 14 | A car travels at 125 km/h for 25 minutes. What distance would it cover?

Distance = Speed × Time
$= 125 \times \frac{25}{60}$
$= 52.1$ km

Change 25 minutes into hours by dividing by 60.

Example 15 | A snail takes 2 hours 24 minutes to climb to the top of a plant of height 35 centimetres. What is its average speed in mm/second?

Speed = Distance ÷ Time = 35 ÷ 2.4 ÷ 60 ÷ 60

2 h 24 min = $2\frac{24}{60}$ hours = 2.4 hours. Divide by 60 and by 60 again to change hours into seconds.

$= 350 \div 2.4 \div 60 \div 60$ Change 35 cm into 350 mm.
$= 0.04$ mm/second

Example 16 | An object travels 25 km in a time of 4 h 45 min. Find its average speed.

Speed = Distance ÷ Time = 25 ÷ 4.75 hours = 5.26 km/h   4 h 45 min = $4\frac{45}{60}$ hours = 4.75 hours

Or we could use the time key on the calculator (see Unit 14). The sequence of key presses is:

25 ÷ 4 [° ' "] 45 [° ' "] [=]

to give the answer 5.26 km/h.

■ Note that some calculators may utilise the second function key to obtain the correct decimal format.

10 | Ratio, proportion, rate

Express direct and inverse variation in algebraic terms and use to find unknown quantities; increase and decrease a quantity by a given ratio

Expressing direct and inverse variation in algebraic terms

The following table shows an example of **direct variation**.

x	0	1	2	3	5	10
y	0	2	4	6	10	20

As the value of x has increased, the value of y has increased also or, reading from right to left, as x decreases, y decreases also.

In algebraic terms we can say: $y = kx$ — y is directly proportional to x. k is the constant of proportionality, i.e. the constant we have to multiply x by to get y.

The following table shows an example of **inverse variation**.

x	1	2	3	4	5	6
y	1	0.5	0.3333 ...	0.25	0.2	0.166 66 ...

This time as the value of x increases, the value of y decreases and vice versa.

In algebraic terms we can say: $y = \dfrac{k}{x}$ — y is inversely proportional to x and again k is the constant of proportionality.

Example 1 y is directly proportional to x. If $y = 6$ when $x = 2$, find:
(a) the constant of proportionality   (b) the value of y when $x = 9$

(a) If y is directly proportional to x then we can say $y = kx$. To find the value of k, we have to rearrange the equation:

$$k = \frac{y}{x} = \frac{6}{2}$$ The given values of x and y are substituted into the equation.

$$k = 3$$

(b) Now that we know the value of k, the y value when $x = 9$ can be found using $y = kx$.

$$y = 3 \times 9$$
$$y = 27$$

Example 2 If y is inversely proportional to x and $y = 4$ when $x = 2.5$, find y when $x = 20$.

First we need to find k, which we can find by rearranging the inverse proportion equation $y = \frac{k}{x}$:

$$k = yx = 4 \times 2.5$$ k, constant of proportionality $= 10$.
$$= 10$$

Now to find y we use the original equation:

$$y = \frac{k}{x} = \frac{10}{20} = 0.5$$ So when $x = 20$, y has a value of 0.5.

| Example 3 | If a stone is dropped off the edge of a cliff, the height (h metres) of the cliff is proportional to the square of the time (t seconds) taken for the stone to reach the ground. A stone takes 5 seconds to reach the ground when dropped off a cliff 125 metres high. |

(a) Write down a relationship between h and t, using k as the constant of proportionality.
(b) Calculate the constant of variation.
(c) Find the height of a cliff if a stone takes 3 seconds to reach the ground.

(a) As h is proportional to t^2 we can say $h = kt^2$.
(b) To find k, we rearrange the equation above and substitute the given values:

$$k = \frac{h}{t^2}$$

$$k = \frac{125}{5^2} = \frac{125}{25} = 5$$

(c) For the height, use the original equation: $h = kt^2$
$$h = 5 \times 3^2 = 5 \times 9 = 45 \text{ metres}$$

Increasing and decreasing a quantity by a given ratio

Sometimes an amount has to be changed according to a given ratio.

| Example 4 | Decrease 60 by the ratio 2 : 3. |

$60 \times \frac{2}{3} = 40$

| Example 5 | A rectangle of 12 cm width and 16 cm length is enlarged in the ratio 3 : 2. Find its new dimensions. |

The rectangle has increased by a factor of $\frac{3}{2}$, so new width $= 12 \text{ cm} \times \frac{3}{2} = 18 \text{ cm}$
new length $= 16 \text{ cm} \times \frac{3}{2} = 24 \text{ cm}$

EXTENDED   Ratio, proportion, rate

11 | Percentages

Calculate a percentage of a quantity; express one quantity as a percentage of another; calculate percentage increase or decrease

Calculating a given percentage of a quantity

Percentage means 'out of a hundred', so when we are required to find a given percentage of an amount, we are really being asked to find a certain 100th of that value.

Example 1

Find 25% of $400.

$$\frac{25}{100} \times \$400 = \frac{25 \times 400}{100} = \$100$$ We are finding 25 hundredths of the $400.

Example 2

A survey was carried out in a school to find the nationality of its students. Of the 220 students in the school, 65% were Kuwaiti, 20% were Egyptian and the rest were European. Find the number of students of each nationality.

Kuwaiti $= \frac{65}{100} \times 220 = 143$

Egyptian $= \frac{20}{100} \times 220 = 44$

European $= \frac{15}{100} \times 220 = 33$

We can check the result by adding each category of student: $143 + 44 + 33 = 220$, our original total.

Example 3

Find the new selling price of a shirt originally costing $20 that has been reduced by 10% in a sale.

Reduction $= \frac{10}{100} \times \$20 = \$2$

New selling price $= \$20 - \$2 = \$18$

The words discount and deduction are also frequently used to describe a reduction. They all mean the same thing, so *subtract* the value.

Alternative method   100% is the whole amount, so a reduction of 10% means that we are now finding 90% of the previous whole amount:

$$90\% \text{ of } \$20 = \frac{90}{100} \times \$20 = \$18$$

Expressing one quantity as a percentage of another

To express one quantity as a percentage of another, first write the first quantity as a fraction of the second and then multiply by 100.

Example 4

In two recent mathematics exams, Lewis scored 56 out of 80 in the first exam and 48 out of 60 in the second exam. In which exam did he score the highest percentage?

First exam $= \frac{56}{80} \times 100 = 70\%$

Second exam $= \frac{48}{60} \times 100 = 80\%$

Therefore his highest score was in the second exam.

Example 5   Four candidates stood in an election:

Candidate A received 24 500 votes.

Candidate B received 18 200 votes.

Candidate C received 16 300 votes.

Candidate D received 12 000 votes.

Express each of these as a percentage of the total votes cast.

Total votes $= 24\,500 + 18\,200 + 16\,300 + 12\,000 = 71\,000$

Candidate A $= \frac{24\,500}{71\,000} \times 100 = 34.5\%$

Candidate B $= \frac{18\,200}{71\,000} \times 100 = 25.6\%$

Candidate C $= \frac{16\,300}{71\,000} \times 100 = 23.0\%$

Candidate D $= \frac{12\,000}{71\,000} \times 100 = 16.9\%$

All answers are given to one decimal place.

Make a simple check by adding all the percentages and ensuring that we get 100% in total, i.e all votes are accounted for.

Calculating percentage increase or decrease

Sometimes the original value and the new value are given and we have to calculate the increase or decrease as a percentage. We can use the formula below to calculate these:

$$\% \text{ Increase/decrease} = \frac{\text{Actual increase/decrease}}{\text{Original value}} \times 100$$

Example 6   A car dealer buys an old car for $1000 and sells it for $1200. Calculate the profit he makes as a percentage increase.

Using the formula, $\% \text{ increase} = \frac{\text{actual increase}}{\text{original value}} \times 100$

we get $\frac{1200 - 1000}{1000} \times 100 = 20\%$ increase

11 | Percentages

Calculate using reverse percentages, e.g. finding the cost price given the selling price and the percentage change

Reverse percentages

This time we are given the selling price and the percentage increase or decrease and we have to work out the cost price. There's a simple formula we can use to work this out:

$$\text{Cost price} = \frac{\text{Selling price}}{\% \text{ change}}$$ If the change was an increase it would be added to 100; if a decrease it would be subtracted.

Example 1   This year a farmer's crop yielded 22 000 tonnes. If this represents a 10% increase on last year, what was the yield last year?

Using the formula $\text{original yield} = \dfrac{\text{new yield}}{110\%}$

we get $\dfrac{22\,000}{110/100}$

$= 20\,000$ tonnes

Note how the cost price is the same as the original yield and the selling price is the same as the new yield. The 100% has been changed to 110% as the yield underwent a percentage increase.

Example 2   A car manufacturer produced 24 000 cars in the month of April, which was a decrease of 4% from the previous month. Calculate the number of cars produced in March.

$\text{Original production} = \dfrac{\text{new production}}{96\%}$

$= \dfrac{24\,000}{96/100}$

$= 25\,000$ cars

This time we have subtracted 4% from 100% as the production suffered a percentage decrease.

■ Note that

- for an **increase** (e.g. profit, tax, surcharge) we **add** the percentage change to 100%.
- for a **decrease** (e.g. reduction, discount, depreciation) we **subtract** the percentage change from 100%.

12 | Using an electronic calculator

Use an electronic calculator efficiently; apply appropriate checks of accuracy

Although there are literally thousands of different types of calculators, with a variety of functions, displays and operating procedures, the most important things to remember when buying your calculator are:

- It should be scientific, i.e. it has all the required functions for IGCSE. Ask your teacher for recommendations.
- It should be straightforward to use, with clear instructions enclosed. Familiarise yourself with the instructions immediately after purchase and practise the examples given.

Even with the huge variety of calculators available, all of the more common types use the same basic keys to find specific answers. These are the most common keys.

Key descriptions

Shift/Inv	This enables the second function on the calculator to be used.
x^2, x^3	Will either square or cube the number input.
x^y	Power key. Will raise the number to the selected power, e.g. 2 $\boxed{x^y}$ 5 = 32.
$\sqrt{\ }, \sqrt[3]{\ }$	Will square or cube root the number.
$\mathbf{a}_{b/c}$	Fraction key, e.g. 3 $\boxed{a\,b/c}$ 5 = 3/5.
Sin, Cos, Tan	Trigonometric functions.
EXP	Standard form. Will express the term as a power of 10, e.g. 4.2 $\boxed{\text{EXP}}$ 3 = $4.2 \times 10^3 = 4200$
MR, Min	Memory keys. Allows numbers to be inserted and recalled from memory.
STO, RCL	Memory keys, found on more modern calculators, allow multiple memory locations.
$\circ\,,\,''$	Time key, allows units of time to be entered. See Unit 14.

Use of the shift or inverse function

Modern calculators are extremely powerful and able to perform far more functions than older models. Many of these extra functions are allocated to existing keys and can be accessed by pressing the shift/inverse button. A good example would be the Sin, Cos, Tan keys, which also hold the Sin^{-1}, Cos^{-1}, Tan^{-1} functions, usually shown above the key in a different colour.

Memory functions such as Min or STO insert number(s) into memory, while MR or RCL recall number(s) from memory. These functions eliminate the need to write the numbers down part way through the calculation, which saves time and avoids rounding errors.

Example 1

Evaluate $\dfrac{34.5 - 28.6}{1.295 \times 3.689\,753}$.

We will use two methods.

Using memory keys $1.295 \times 3.689\,753$ $\boxed{\text{STO A}}$ $\boxed{\text{AC}}$ $34.5 - 28.6 = \div$ $\boxed{\text{RCL}}$ $\boxed{\text{A}}$
$= 1.235$ (to 3 d.p.)

Using bracket keys $\boxed{(}$ $34.5 - 28.6$ $\boxed{)}$ $\div$ $\boxed{(}$ $1.295 \times 3.689\,753$ $\boxed{)}$ $= 1.235$ (to 3 d.p.)

Example 2

Find $\dfrac{4^3 \times 2^5 \times \sqrt{16}}{3.2 \times 10^4 + \frac{3}{4}\pi}$.   Note that both standard form and π (2nd function) are required for this calculation.

It will be quicker to use brackets again. Note the key selections:

$\boxed{(}$ 4 $\boxed{x^3}$ $\times 2$ $\boxed{x^y}$ $5 \times$ $\boxed{\sqrt{}}$ 16 $\boxed{)}$ $\div$ $\boxed{(}$ 3.2 $\boxed{\text{EXP}}$ $4 + 3$ $\boxed{a\,b/c}$ $4 \times$ $\boxed{\text{Shift}}$ $\boxed{\pi}$ $\boxed{)}$
$= 0.256$ (to 3 d.p.)

Example 3

Find $\text{Sin}^{-1} -0.6^2$. Sin^{-1} is the second function, and the 0.6 can be made negative by using the $\boxed{(-)}$ key.

$\boxed{\text{Shift}}$ $\boxed{\text{Sin}^{-1}}$ $\boxed{(}$ $\boxed{(-)}$ $\boxed{0.6}$ $\boxed{)}$ $\boxed{x^2}$ $= 21.1°$

■ Note that the $\boxed{(-)}$ is often a source of incorrect answers. If the negative term is raised to a power, then the bracket key needs to be used before and after the term. Otherwise the calculator will raise the term to the power and then make the result negative.

13 | Measures

Know and be able to convert units of mass, length, area, volume and capacity in practical situations

Units of mass, length and capacity

Below are the standard units of mass, length and capacity, together with their conversions to smaller or larger units.

Mass (weight)

The units of mass are the tonne (t), kilogram (kg), gram (g) and milligram (mg).

The conversions are as follows:
$$
\begin{aligned}
1 \text{ tonne} &= 1000 \text{ kilograms} \\
1 \text{ kilogram} &= 1000 \text{ grams} \\
1 \text{ gram} &= 1000 \text{ milligrams}
\end{aligned}
$$

This flow diagram shows the conversions.

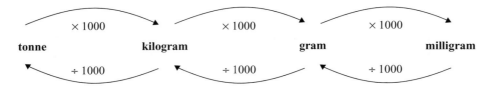

When you change, for example, tonnes to kilograms, you multiply the tonnes figure by 1000 to get the equivalent kilograms figure. To change kilograms to tonnes you divide by 1000.

Length

The units of length are the kilometre (km), metre (m), centimetre (cm) and millimetre (mm).

The conversions are:
$$
\begin{aligned}
1 \text{ kilometre} &= 1000 \text{ metres} \\
1 \text{ metre} &= 100 \text{ centimetres} \\
1 \text{ centimetre} &= 10 \text{ millimetres}
\end{aligned}
$$

This flow diagram shows the conversions.

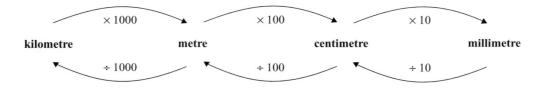

For example, to change kilometres to metres you multiply the kilometre figure by 1000 to get the equivalent metres figure. To change metres to kilometres you divide by 1000.

Capacity

The word capacity is used to measure liquid volumes. The units of capacity are the litre (l), centilitre (cl) and millilitre (ml).

The conversions are as follows: 1 litre   = 100 centilitres

1 centilitre = 10 millilitres

This flow diagram shows the conversions.

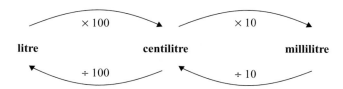

When changing litres to centilitres you multiply the litres figure by 100 to get the equivalent centilitres figure. To change centilitres to litres you divide by 100.

Example 1   Convert each of the following lengths to metres:

(a) 2.8 km   (b) 672 cm   (c) 28 400 mm

Using the flow diagram:

(a) km $\longrightarrow$ metres = × 1000   2.8 km = 2.8 × 1000   = 2800 metres

(b) cm $\longrightarrow$ metres = ÷ 100   672 cm = 672 ÷ 100   = 6.72 metres

(c) mm $\longrightarrow$ metres = ÷ 10 ÷ 100   28 400 mm = 28 400 ÷ 10 ÷ 100 = 28.4 metres

---

Example 2   How many 25 ml doses can be obtained from a bottle containing 1.7 litres of medicine?

In a problem where the two numbers are in different units, we must first change one of them to the other unit:

litres $\longrightarrow$ ml = ×10 × 100   1.7 litres = 1.7 × 10 × 100 = 1700 ml

Number of 25 ml doses = 1700 ÷ 25 = 68

Example 3   15.4 tonnes of a medicine are to be made into tablets. If each tablet weighs 7 mg, calculate the number of tablets that can be made.

The units are different, so change the 15.4 tonnes into mg:

tonnes $\longrightarrow$ mg = 15.4 × 1000 × 1000 × 1000 = 15 400 000 000 mg

Number of tablets = 15 400 000 000 ÷ 7 = 2 200 000 000

■   Note that we could have changed the 7 mg into tonnes and then divided as before to get the same answer.

Area and volume

Calculations involving areas and volumes of different shapes are looked at in detail in Unit 31, but you should be able to recognise which formulae refer to area and which refer to volume:

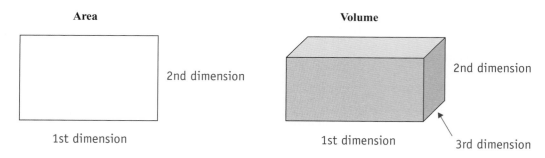

As **area** involves two dimensions multiplied together, e.g. length × width for a rectangle, the answer is always given as units squared (e.g. cm^2, mm^2).

With **volume** there are three dimensions multiplied together, e.g. length × width × height for a cuboid, so the answer is always given as units cubed (e.g. cm^3, m^3).

Here are some useful area and volume formulae:

Area (involves 2 dimensions)		Volume (involves 3 dimensions)	
$\frac{1}{2}bh$	Area of a triangle	$\frac{4}{3}\pi r^3$	Volume of a sphere
$2\pi r^2 + 2\pi rh$	Surface area of a cylinder	$\frac{1}{3}\pi r^2 h$	Volume of a cone
$4\pi r^2$	Surface area of a sphere	$\pi r^2 h$	Volume of a cylinder
$\frac{1}{2}(a+b)h$	Area of a trapezium		

14 | Time

Understand time in terms of the 24-hour and 12-hour clock; read clocks, dials and timetables

The 12- and 24-hour clock

There are two ways of showing the time, by using either the 12- or 24-hour clock. With the 12-hour clock there are two periods, each of twelve hours' duration, during each day. The period between midnight and noon is called a.m. (*ante meridiem*, meaning before midday) and the period between noon and midnight is called p.m. (*post meridiem*, meaning after midday).

With the 24-hour clock there is one twenty-four hour period starting at midnight and finishing at midnight the following day. This type of clock system is generally used for timetables for buses and trains and is always given as **four figures**.

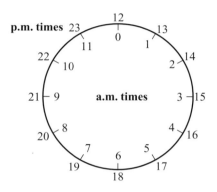

The clock shows that once the first 12-hour duration has finished (the a.m. times), then the 24-hour time continues, using 13 to represent 1 o'clock p.m. and 14 to represent 2 o'clock p.m. and so on until it completes its cycle of 24 hours.

For example, 2.30 p.m. = 1430 hours
8.45 p.m. = 2045 hours

Changing 12-hour time to 24-hour time

Both a.m. time and the 24-hour clock time start simultaneously at midnight. To convert, we simply replace the a.m. with hours or (h) and insert a zero before the digits, as necessary, to make it four figures:

Example 1	Change 4.25 a.m. and 10.45 a.m. into 24-hour clock times.
4.25 a.m. = 0425 hours	Remember to add zero.
10.45 a.m. = 1045 hours	No need to add zero – the a.m. time already has four digits.

When we change a p.m. time to 24-hour clock time, we must take into account the previous 12-hour cycle.

Example 2	Change 8.50 p.m. into 24-hour clock time.
8.50 p.m. = 0850 hours + 12 hours = 2050 hours Add 12 hours for the previous a.m. cycle.	

Changing 24-hour time to 12-hour time

This is simply a reversal of the previous method. If we have a time greater than 1200 hours, then it must have travelled through the a.m. time period. So we subtract 1200 hours to get the correct 12-hour p.m. time.

Example 3 Change 0858 hours and 1835 hours to 12-hour clock time.

$$0858 \text{ hours} = 8.58 \text{ a.m.}$$

Must be a.m. since it is less than 1200 hours.

$$1835 \text{ hours} = 1835 - 1200$$
$$= 6.35 \text{ p.m.}$$

Subtract the first cycle of 12 hours.

Here are some more examples:

12 hour	24 hour
4.25 a.m.	0425 hours
4.25 p.m.	1625 hours
8.15 a.m.	0815 hours
9.35 p.m.	2135 hours

Calculating with time

We will deal with the calculator method on page 41, but you should also be able to perform the necessary steps without using a calculator.

Remember the conversions:

1 day = 24 hours   1 hour = 60 minutes   1 minute = 60 seconds

Example 4 Add 3 hours 25 minutes to 2 hours 45 minutes.

First set out the calculation as shown:

Hours	Minutes
3	25
2	45 +
5	70
↓	
6	10

We can't leave the time as 5 hours 70 minutes so we have to carry one hour, i.e. 60 minutes, over to the hours column. It then becomes 6 hours 10 minutes.

Example 5 Find the time difference between 10.45 a.m. and 8.34 p.m.

First change both times to 24-hour clock time: 10.45 a.m. = 1045 hours
8.34 p.m. = 2034 hours

Now subtract the smaller time from the larger time:

Hours		Minutes	
20		34	
−10		45	
(20 − 1)		(34 + 60)	
19	94		
10	45		
9	49		

The minutes column can't be subtracted normally because 45 is bigger than 34, so we have to borrow one hour from the 20 and add it to the 34 minutes. It now becomes 1994 hours. Then we subtract as normal.

■ Note that occasionally the minutes will subtract normally without having to borrow an hour.

Calculator method $\boxed{\circ\,{'}\,{''}}$

The key shown above enables calculations with hours, minutes and seconds to be done. The previous example could have been done by pressing the following keys on your calculator

20 $\boxed{\circ\,{'}\,{''}}$ 34 $\boxed{\circ\,{'}\,{''}}$ − 10 $\boxed{\circ\,{'}\,{''}}$ 45 $\boxed{\circ\,{'}\,{''}}$ $\boxed{=}$

to give the answer 9° 49' 0".

This is the same as 9 hours, 49 minutes and 0 seconds.

Timetables

This is a timetable of the Cheltenham to Teddington bus, which runs from Monday to Saturday weekly. (All times shown in the table are in 24-hour clock time.)

	1st bus ↓	2nd bus ↓	3rd bus ↓						Starting point ↓
Cheltenham	0705	0735	0830	0930	0955	1020	1125	1200	1230
Cleeve Estate	0717	0747	0842	0942	1007	1032	1137	1212	1242
Bishops Cleeve	0725	0755	0850	0950	1015	1040	1145	1220	1250
Gretton	0735	-	-	-	-	1050	-	-	-
Alstone	-	-	0909	-	-	-	-	-	-
Teddington	-	-	0914	-	-	-	-	-	-
Cheltenham	1255	1335	1455	1455	1605	1700	1730	1800	1930
Cleeve Estate	1307	1347	1507	1507	1617	1712	1742	1812	1942
Bishops Cleeve	1315	1355	1515	1515	1625	1720	1750	1820	1850
Gretton	1325	1405	-	1525	1635	1730	1800	1830	-
Alstone	1340	-	-	-	1645	-	-	1840	-
Teddington	1345	-	-	-	1650	-	-	1845	-

↑ Finishing point

From the timetable we can see that the first bus commences at 0705 hours, continues through the towns of Cleeve Estate and Bishops Cleeve, and finally reaches Gretton at 0735 hours. The second bus leaves Cheltenham at 0735 hours and finishes at Bishops Cleeve at 0755, and so on.

We also notice that some buses call at certain villages, while others do not, so this must be kept in mind when planning a journey. Now for a few examples.

| Example 6 | What time does the 0930 bus from Cheltenham arrive in Bishops Cleeve? Calculate its journey time.

By inspection of the table, the 0930 is the fourth bus and arrives at 0950 in Bishop's Cleeve.

To calculate the journey time, simply subtract the start time from the finishing time:

0950 − 0930 = 20 minutes |

Example 7 I want to travel from Bishops Cleeve to Alstone.

(a) How long does the 1315 bus take for the journey?

(b) How long does the 1625 bus take?

(c) What is the difference in the time taken?

(a) From the table we can see that the 1315 bus from Bishops Cleeve is the 10th bus and it arrives in Alstone at 1340. By subtracting the times we find:

$$1340 - 1315 = 25 \text{ minutes}$$

(b) From the table the 1625 bus arrives at 1645, which is a journey of:

$$1645 - 1625 = 20 \text{ minutes}$$

(c) Difference in time taken $= 25 - 20$

$$= 5 \text{ minutes}$$

Reading dials

The diagram below shows the dials of an electricity meter, each dial showing its actual numerical value:

dial 1 $= 5 \times 10\,000$

dial 2 $= 6 \times 1000$

dial 3 $= 3 \times 100$

dial 4 $= 7 \times 10$

dial 5 $= 8 \times 1$

Therefore the total value of electricity used to date is 56 378 kWh.

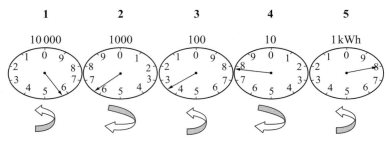

■ Note that we read the dials alternately anti-clockwise, then clockwise. Remember that if the pointer comes between two numbers, choose the lower number for the actual value.

15 | Money

Calculate using money and convert from one currency to another

Exchange rates

Every country has its own monetary system, for example Japan has the yen, Britain the pound, America the dollar. If there is to be trade and travel between any two countries there has to be a rate at which the money of one country can be converted into the money of another country. This is called the rate of exchange. Below is a table showing some typical exchange rates.

Country	Rate of exchange
Australia	1.65 dollars
Cyprus	0.58 pounds
Sweden	8.93 kronor
Eurozone	1.03 euros
Japan	116.34 yen
Thailand	36.37 bahts
Canada	1.37 dollars
Switzerland	1.44 Sfrancs

US$ foreign exchange rates at January 2003

The table shows exchange rates for the American dollar $. For every $1 we get 1.03 euros, for $2 we get 2.06 euros and so on. Here are a few examples using the above exchange rates.

Example 1 Change $300 into Japanese yen.

dollars $1   : $300 $\dfrac{x}{300} = \dfrac{116.34}{1}$

yen   116.34 : x $x = \dfrac{300 \times 116.34}{1} = 34\,902$ yen

Example 2 Change 2760 bahts into American dollars.

dollars $1   : x $\dfrac{x}{2760} = \dfrac{1}{36.37}$

bahts   36.37 : 2760 $x = \dfrac{2760 \times 1}{36.37} = \$75.89$

■ Note that the accuracy to which you give your answer should be appropriate for the currency you're dealing in. For example, answers in US dollars and UK pounds need only be given to two decimal places since the smallest denominations available are cents and pence respectively (2nd decimal place).

16 | Personal and household finance

Personal and household finance involving earnings, simple interest, discount, profit and loss; extract data from tables and charts

Personal and household finance

There are many words related to personal and household finance. The more common ones are listed below.

Salary	Payment made to a person for doing a period of work. Salaries are usually paid monthly.
Wages	Again, payment made to a person for doing a period of work, but usually paid weekly.
Overtime	Extra hours of work, for which the person is usually paid an increased amount.
Piecework	Where the person is paid a fixed amount for each article or piece of work that they make.
Commission	Where the worker is paid a percentage of the total amount of items sold. Salesmen, shop assistants and representatives are usually paid on commission.
Income tax	A tax levied by a government on salaries to help pay for the country's services. It is usually a percentage figure on an incremented scale depending upon how much the worker earns.
Pension fund	A percentage of the employee's salary is paid into a pension fund in order to provide income after retirement.
Profit	A profit is made when the selling price is greater than the cost price. (You make money.)
Loss	A loss is made when the selling price is less than the cost price. (You lose money.)
Discount	A reduction of an amount, usually given as a number or as a percentage figure.

Here are some examples on a few of the above terms.

Example 1 An employee in a clothes factory earns $200 for a 40-hour week. Find:
(a) his hourly rate   (b) his overtime rate if overtime is paid at time and a half.

(a) Hourly rate $= \$200 \div 40 = \$5$ per hour
(b) Overtime rate $= \$5 \times 1\frac{1}{2} = \$7.50$ per hour

---

Example 2 A woman earns $8500 per annum. She pays 9% of her salary in pension contributions. How much does she pay per annum?

Amount paid $= \frac{9}{100} \times \$8500 = \$765$

Example 3 A man's salary is $14 000 per annum and his total tax-free allowances are $4500.
(a) Work out his taxable income.
(b) If income tax is levied at 25%, calculate the amount of tax payable.

(a) Taxable income is the amount of earnings he must pay tax on after his tax-free allowances have been deducted.

Taxable income $= \$14\,000 - \$4500 = \$9500$

(b) Tax payable $= \frac{25}{100} \times \$9500 = \$2375$

■ For examples on profit, loss and discount, see Unit 11.

Simple interest

When money is invested with a bank, interest is paid to the investor for lending the money. The amount of interest paid depends on:

- the **principal** amount invested
- the percentage **rate** of interest
- the **time** period

The simple interest formula is:

$$I = \frac{PTR}{100} \qquad \begin{array}{ll} I = \text{Interest} & P = \text{Principal} \\ T = \text{Time} & R = \text{Rate}\,(\%) \end{array}$$

Example 4 Find the interest payable on $500 invested for 2 years at 6% per annum.

Using the formula: $I = \dfrac{PTR}{100} = \dfrac{500 \times 2 \times 6}{100} = \$60$

Example 5 Find the rate in per cent per annum of simple interest if $1200 is the interest on $3000 invested for 4 years.

Rearranging the formula: $R = \dfrac{100I}{PT} = \dfrac{100 \times 1200}{3000 \times 4} = 10\%$

Extracting data from tables and charts

There are countless types of charts and tables, many of which are looked at elsewhere, e.g. timetables in Unit 14 and bar charts in Unit 33. Here are two more examples.

A table showing compound interest paid by a bank to an investor

Years	5%	6%	7%	8%	9%
1	1.050	1.060	1.070	1.080	1.090
2	1.103	1.124	1.145	1.166	1.188
3	1.158	1.191	1.225	1.260	1.295
4	1.216	1.262	1.311	1.360	1.412
5	1.276	1.338	1.403	1.469	1.539
6	1.340	1.419	1.501	1.587	1.677
7	1.407	1.504	1.606	1.714	1.828
8	1.477	1.594	1.718	1.851	1.993
9	1.551	1.689	1.838	1.999	2.172
10	1.629	1.791	1.967	2.159	2.367

The years column is on the left and the percentage rate is on the top row. The figures in the table are multiplied by the principal amount invested to give the total money returned at the end of the time period.

Example 6 If the principal amount is $2000, the time period is 4 years and the rate of interest is 6%, what is the total money returned?

From the table, the multiplying factor = 1.262 (shaded box), so total money returned = 1.262 × $2000 = $2524.

A pie chart showing the total sales of a department store for one day

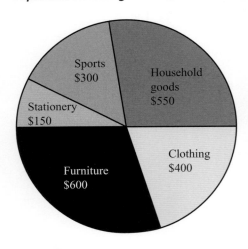

The pie chart shows that the highest sales were in the furniture section as this has the largest sector in the chart, followed by household goods, clothing, sports equipment and stationery.

Example 7	Find as a percentage the amount of sports equipment sold compared to the total sales.

Sports equipment = $300

Total sales = $2000

$$\% \, \text{sports} = \frac{300}{2000} \times 100$$

$$= 15\%$$

17 | Graphs in practical situations

Interpret and use graphs in practical situations including travel graphs and conversion graphs; draw graphs from given data

Graphs with two dimensions

Graphs with two dimensions have two axes which allow you to compare two variables; the line of the graph shows you the relationship between the variables. Here are some graphs with two dimensions.

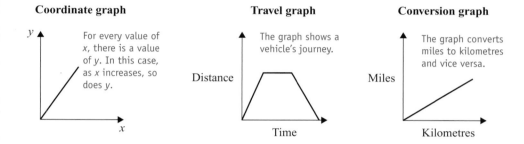

Coordinate graph
For every value of x, there is a value of y. In this case, as x increases, so does y.

Travel graph
The graph shows a vehicle's journey.
Distance / Time

Conversion graph
The graph converts miles to kilometres and vice versa.
Miles / Kilometres

Interpreting and using graphs in practical situations

The two most common practical graphs are:

- **travel graphs**, where the vertical axis shows the distance travelled and the horizontal axis shows the time taken for each part of the journey
- **conversion graphs**, where the vertical axis shows one variable and the horizontal axis shows the other variable, and the line of the graph gives the conversion

Travel graphs

The graph below shows a train's journey from Granada to Madrid, a distance of 200 miles.

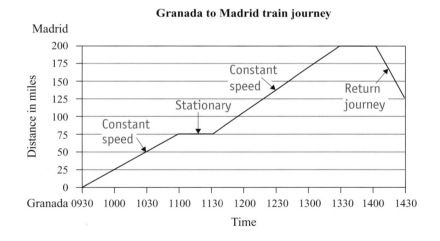

Granada to Madrid train journey

We can draw the following conclusions from the graph:

- The train left Granada at 0930 hours.
- It continued at a constant speed until 1100 hours (straight line graph).
- It was stationary from 1100 to 1130 hours.
- It continued at a constant speed again from 1130 to arrive in Madrid at 1330 hours.
- The return journey began at 1400 hours back to Granada.

Conversion graphs

The graph below shows the conversion of ounces (imperial weight) to grams (metric weight).

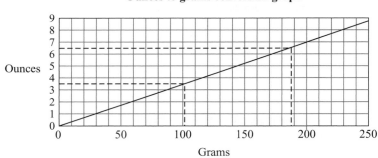

Ounces to grams conversion graph

We have a constant (linear) graph, e.g.

> 1 ounce ≈ 28 grams
> 2 ounces ≈ 56 grams

The number of grams is always, approximately, the ounces figure multiplied by 28.

We can use the graph to find any corresponding value of ounces or grams, e.g.

3.5 ounces = approx. 100 grams   6.5 ounces = approx. 188 grams

Drawing graphs from given data

When drawing the graph, bear in mind the following guidelines:

- **Decide which axis is to show which variable**. For conversion graphs it doesn't really matter since it will be a straight line graph anyway. However, for distance–time graphs we normally put distance on the vertical axis and time on the horizontal axis.
- **Select a suitable scale** for your graph. The graph should cover all required points and be neither too large (too much detail) or too small (not enough detail).
- **Label your axes**, e.g. distance on the vertical axis and time on the horizontal axis.
- Give your graph a **title**.

Example 1

Given that 80 km = 50 miles, draw a conversion graph up to 80 km. Use your graph to estimate:
(a) how many miles is 50 km   (b) the speed in km/h equivalent to 40 mph.

We only need two points to be able to draw our straight line graph. We have the final point, 80 km = 50 miles, so we need a starting point.
By calculation, $10 \text{ miles} = \frac{80}{50} \times 10 = 16 \text{ km}$.

As it is a conversion graph it doesn't matter which axis we use for miles or km, so let's make the vertical axis the kilometres variable and the horizontal axis the miles variable.

Remember that the graph should be drawn on graph paper so that the points can be plotted more accurately. We then label the axes and give the graph a title.

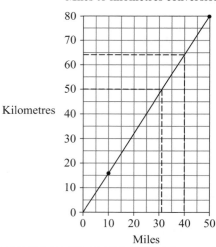

Miles to kilometres conversion

(a) Using the graph, 50 km = approx. 32 miles.

(b) As speed is consistent with the graph,
    40 mph = approx. 64 km/h.

---

17 | Graphs in practical situations

Apply rate of change to speed–time graphs, including calculating distance travelled as area under a linear speed–time graph; calculate acceleration and deceleration

Speed–time graphs

This time the vertical axis shows the speed and the horizontal axis shows the time. Below is an example with the relevant parts explained.

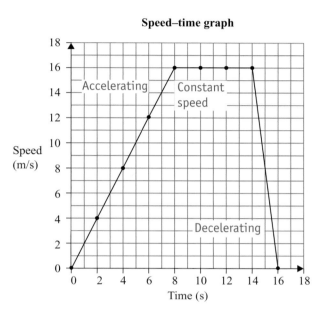

Speed–time graph

Accelerating = increase in speed
Decelerating = decrease in speed
Constant speed = no change in speed

From the graph we can draw the following conclusions:

- From 0 seconds (rest) to 8 seconds the vehicle increases its speed from 0 m/s to 16 m/s at a constant acceleration.
- From 8 seconds to 14 seconds the vehicle maintains a constant speed of 16 m/s (no acceleration).
- From 14 seconds to 16 seconds the vehicle decreases its speed from 16 m/s to 0 m/s (back to rest) at a constant deceleration.

Summary

- A straight horizontal line represents constant speed.
- A positive gradient indicates an acceleration; a negative gradient indicates deceleration.
- The steepness of the gradient indicates the rate of acceleration/deceleration. (The steeper the slope, the faster the increase/decrease in speed.)

Calculating the distance travelled from the area under a speed–time graph

The area under a speed–time graph gives the total distance travelled.

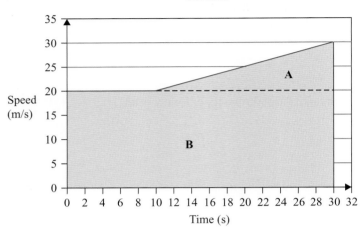

Calculating distance travelled

The blue line shows the speed–time graph.

In order to calculate the area under the graph we have to split it into two distinct shapes:

Shape A is a triangle.
Shape B is a rectangle.

Now we can work out the areas of each of the shapes and add them to get the total area:

$$\text{Area A} = \frac{\text{base} \times \text{height}}{2}$$

$$= \frac{20 \times 10}{2} = 100\,\text{m}$$

$$\text{Area B} = \text{length} \times \text{width}$$

$$= 30 \times 20 = 600\,\text{m}$$

$$\text{Total area} = 100\,\text{m} + 600\,\text{m} = 700\,\text{m}$$

Therefore the total distance travelled is 700 metres.

Calculating the acceleration and deceleration

We can calculate both of these using the formula:

$$\text{Acceleration/deceleration} = \frac{\text{Change in speed}}{\text{Time taken}}$$

So if we wanted to calculate the acceleration for the second part of the journey above, i.e. from 10 seconds to 30 seconds:

$$\text{Acceleration} = \frac{\text{Change in speed}}{\text{Time taken}} = \frac{30 - 20}{30 - 10} = \frac{10}{20} = 0.5\,\text{m/s}^2$$

■ Note that we know that we are calculating **acceleration** since the graph shows a **positive gradient**. It is possible to get the same answer by splitting the area under the speed–time graph into a different arrangement of shapes, provided we know the formulae for the areas of these shapes (e.g. a trapezium or a square).

18 | Graphs of functions

Construct tables of values for linear, quadratic and reciprocal functions; draw and interpret such graphs; find the gradient of a straight line graph; solve linear and quadratic equations approximately by graphical methods

Linear graphs of the form $y = ax + b$ (where a and b are integers)

These type of graphs are straight line graphs. Any function which involves x^1, or in other words just x, must be a straight line graph.

Example 1 | Draw the graph of the function $y = 2x + 1$, using values of x from -3 to $+3$.

Step 1 Construct a table using the x-values given.

x	-3	-2	-1	0	1	2	3
$y = 2x + 1$	-5	-3	-1	1	3	5	7

When $x = -3$
$y = 2(-3) + 1$
$= -5$

When $x = -1$
$y = 2(-1) + 1$
$= -1$

When $x = 1$
$y = 2(1) + 1$
$= 3$

Step 2 Plot the graph, using the values from the table. Looking at the table, the x-values range from -3 to $+3$ and the y-values range from -5 to $+7$.

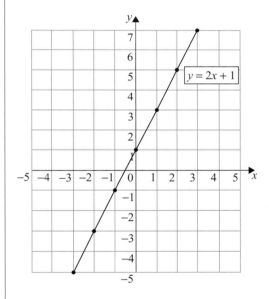

■ Note that all points fit exactly on the line. Any points that were not on the line should be checked for either errors in their calculations or misplotting.

Quadratic graphs of the form $y = \pm x^2$

These type of graphs are curved or parabolic graphs. Any graph which involves x^n where n is greater than 1 is a curved graph. The example involves the same basic steps as before.

Example 2 Draw the graph of x^2 using values of x from -3 to $+3$.

Step 1 Construct a table using the x-values given.

x	-3	-2	-1	0	1	2	3
$y = x^2$	9	4	1	0	1	4	9

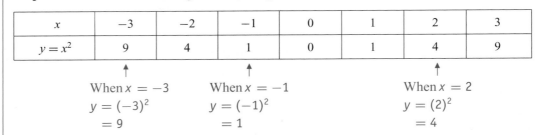

When $x = -3$ When $x = -1$ When $x = 2$
$y = (-3)^2$ $y = (-1)^2$ $y = (2)^2$
$= 9$ $= 1$ $= 4$

Step 2 Plot the graph using the x- and y-values from the table.

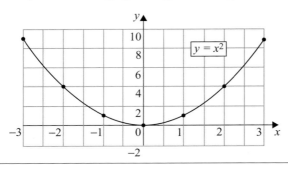

Example 3 Draw the graph of $-x^2$ using x values of -3 to $+3$.

Draw the table and hence plot the graph:

x	-3	-2	1	0	1	2	3
$y = -x^2$	-9	-4	-1	0	-1	-4	-9

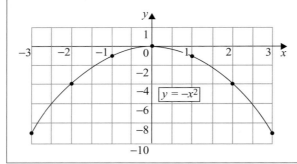

We can see from the graph that the shape is similar to the graph of $y = x^2$. But it is upside down, so the effect of the negative sign is to invert the graph.

Use of function notation in graphs $f(x) = ax + b$

Very often the notation $f(x) = ax + b$ is used, instead of $y = ax + b$. So we are now calculating the $f(x)$ value for varying values of x instead of the y-values. In effect the vertical axis will now represent $f(x)$, rather than y.

The method for constructing and solving these type of questions is the same as before.

Example 4   Draw the graph of $f(x) = 2x^2 + 3x - 1$, using x-values of -3 to $+1$.

Draw the table and put in the y-values.

x	-3	-2	-1	0	1
$2x^2$	18	8	2	0	2
$3x$	-9	-6	-3	0	3
-1	-1	-1	-1	-1	-1
$f(x) = 2x^2 + 3x - 1$	8	1	-2	-1	4

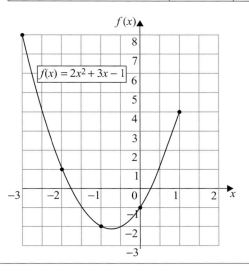

■ Note that the shape of the graph is very similar to the x^2 graph but it has been shifted to the left. The other values that have been added to the x^2 term are responsible for this.

Reciprocal graphs of the form $y = \frac{a}{x}$ (where $x \neq 0$)

Reciprocal graphs are also curved graphs. The final shape of the graph depends on the values applied to x. They are sometimes symmetrical graphs as the next example shows.

Example 5   Draw the graph of $y = \frac{5}{x}$ using values of x from -3 to $+3$.

Draw the table and calculate the y-values.

x	-3	-2	-1	0	1	2	3
$y = \frac{5}{x}$	-1.67	-2.5	-5	$-$	5	2.5	1.67

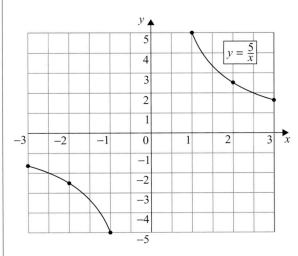

■ Note that we have a symmetrical graph, the line of symmetry being the line $y = -x$. The ends of the curves will never actually touch the x- or y-axes because $1 \div 0 = $ infinity.

The process is the same for any $\frac{a}{x}$ graph although the final shape of the graph will alter according to the actual function values.

The gradient of a graph

The gradient is the slope of the graph. It tells us how much the graph rises or falls for each x unit change. The gradient can be positive or negative depending on which direction the graph is sloping and it can be steep or shallow depending on the function.

The general formula for finding the gradient is:

$$\text{Gradient} = \frac{\text{change in } y}{\text{change in } x} \quad \text{where } x \text{ is the horizontal size and } y \text{ the vertical size}$$

Example 6	The line AB has coordinates A(2, 3) and B(4, 9). Find its gradient.

By use of a sketch (no need to scale).

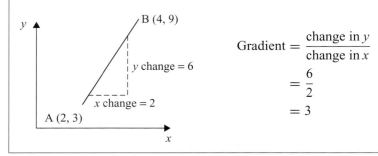

$$\text{Gradient} = \frac{\text{change in } y}{\text{change in } x}$$
$$= \frac{6}{2}$$
$$= 3$$

Types of gradients

A positive steep gradient

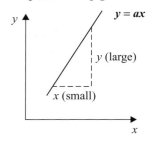

A positive shallow gradient

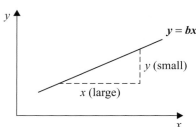

A negative steep gradient

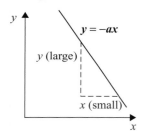

A negative shallow gradient

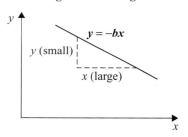

Finding the gradient of a graph

In order to find the gradient of any straight line graph, we choose any point along its length, form a right-angled triangle and simply read off the corresponding changes in x and y. Then we use the formula for the gradient, gradient = (change in y) ÷ (change in x).

Example 7 Find the gradient for the graph of $y = 2x + 1$.

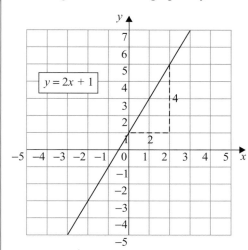

Step 1 We start the triangle at coordinates (0, 1) and go across 2 units in the x-direction. Then we have to go up for 4 units in order to touch the graph again.

Step 2 Using the formula:

$$\text{Gradient} = \text{change in } y \div \text{change in } x$$
$$= 4 \div 2$$
$$= 2$$

■ Now we have found the gradient to be 2 for the graph of $y = 2x + 1$, we know that for every one unit the graph moves horizontally, it will move two units vertically. It also tells us that it is a positive gradient in the x-direction, i.e. both x and y increase in value when moving from left to right. If there was a negative sign in front of the 2, then the gradient would have been negative.

Example 8 Find the gradient for the graph of $y = -3x + 2$.

First we construct our table of values and plot the graph.

x	-2	-1	0	1	2
$y = -3x + 2$	8	5	2	-1	-4

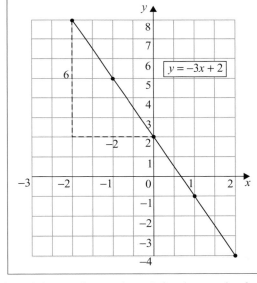

Step 1 We start the triangle at the point (0, 2) and go across -2 units in the x-direction. Then we have to go up 6 units in order to touch the graph again.

Step 2 Using the formula:

$$\text{Gradient} = y \div x$$
$$= 6 \div -2$$
$$= -3$$

■ We have found the gradient to be -3 for the graph of $y = -3x + 2$. This tells us that for every one unit in the x-direction the graph moves horizontally, it will move three units vertically. It also tells us that it is a negative gradient, i.e. x and y vary inversely in value when moving from left to right (as x increases, y decreases and vice versa).

Solving linear equations by graphical methods $y = ax + b$

We know that any equation of the form x^1 or simply x is linear and that it is a straight line graph.
Once you have drawn the graph of an equation, it is very easy to solve it by simple observation of
the graph.

Example 9   Use the graph of $y = x + 3$ to solve the equation $0 = x + 3$.
Use x-values of -3 to $+3$ and check that the solution is correct.

In effect the question is asking for the x-value that will give $y = 0$.

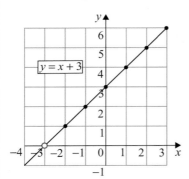

$y = 0$ is the x-axis and we can see that the graph crosses the
x-axis at $x = -3$, the point circled on the graph.

Therefore the solution for the graph

$$0 = x + 3 \text{ is } x = -3.$$

Now we can check.

Substituting our $x = -3$ value from the graph:

$$x + 3 = 0$$
$$(-3) + 3 = 0$$
$$0 = 0$$

Both sides of the equation balance, therefore the solution is correct.

Solving quadratic equations ($y = ax^2 + b$) by graphical methods

We know that graphs involving x^2 terms are curved or parabolic in shape. The index value above
the x also tells us how many possible solutions of x there are for the equation, so in this case there
must be two solutions for x. The method of solving quadratic equations using graphs is the same as
for linear equations, except that the graph will now have two solutions for x. (The graph will
intersect the appropriate line in two places not just one.)

Example 10   Solve the quadratic function $-3 = x^2 - 4$ by using a graphical method, using x-values of -3 to
$+3$. Check that the solution is correct using any other method.

We will draw the graph of $y = x^2 - 4$ and then, by observation of the graph, find where it
intersects the line $y = -3$.

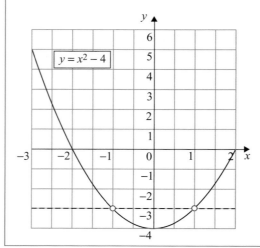

The dotted line $y = -3$ intersects the graph of
$y = x^2 - 4$ at two points (circled):

$$x = 1 \quad \text{and} \quad x = -1$$

Therefore the solutions for the graph are:

$$x = 1 \text{ and } x = -1$$

Now we can check.
Substituting our values of $x = -1, 1$ from the graph:

$$(-1)^2 - 4 = -3$$
$$\text{and } (1)^2 - 4 = -3$$

In both cases, both sides of equation balance, so the
solutions are correct.

18 | Graphs of functions

Construct tables of values and draw graphs of quadratic, cubic and reciprocal functions; estimate gradients of curves by drawing tangents; solve associated equations approximately by graphical methods

Quadratic, cubic and reciprocal graphs

There are countless forms that these graphs can take but the method of constructing the table and then plotting the graph is essentially the same as for the work covered in the core section. We will look at two more examples to illustrate this.

Example 1 | Draw the graph of the function $y = (3 + 2x)(3 - x)$ for values of x from –2 to 4. Use your graph to solve $0 = (3 + 2x)(3 - x)$.

Construct the table and plot the graph.

x	–2	–1	0	1	2	3	4
$y = (3 + 2x)(3 - x)$	–5	4	9	10	7	0	–11

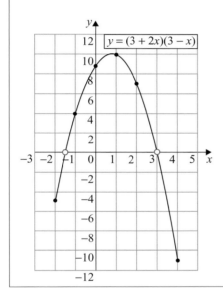

Remember that a $-x^2$ graph should be an inverted parabola and that the graph should cut the x-axis in two places in order to obtain the two solutions.

We can see from the graph again that it does in fact have two solutions,

$$x = -1.5 \quad \text{and} \quad x = 3,$$

and that it is an inverted parabola.

Finding the gradient of a curved graph

Because a curved graph is constantly changing gradients, both in size and direction, we need to know at which point on the graph to measure the gradient. We then draw a tangent to this point and use it to construct our right-angled triangle. When we have this, then it is a simple matter of using the formula to calculate the gradient.

Example 2

The following table shows some values for the function $y = \frac{x^3}{12} - \frac{6}{x}$.

Values of y are given to one decimal place.

x	0.6	1	1.5	2	2.5	3	3.5	4	4.5	5
y	p	−5.9	−3.7	−2.3	−1.1	0.3	1.9	3.8	q	r

(a) Find the values of p, q and r.   (b) Use the values in your table to draw the graph.

(c) Find, from your graph, correct to one decimal place, the value of x for which $\frac{x^3}{12} - \frac{6}{x} = 0$.

(d) Draw the tangent to the curve at the point where $x = 1$ and hence estimate the gradient of the curve at that point.

(a) $p = (0.6)^3/12 - 6/0.6 = 0.018 - 10 = -9.9$ By simple substitution of the x-values into
$q = (4.5)^3/12 - 6/4.5 = 7.594 - 1.33\ldots = 6.3$ the function.
$r = (5)^3/12 - 6/5 = 10.4 - 1.2 = 9.2$

(b) Now we plot the graph:

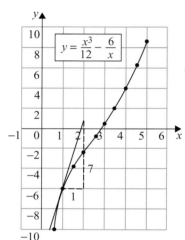

(c) We see from the graph that it crosses the x-axis (where $y = 0$) at the point $x = 2.9$, so this is the solution.

(d) Draw the tangent to the graph at $x = 1$ and construct the triangle, read off the x- and y-coordinates and find the gradient.

$$\text{Gradient} = \frac{\text{change in } y}{\text{change in } x} = \frac{7}{1}$$
$$= 7$$

■ Note that the changes in x and y are approximate and have been rounded to the nearest whole number.

19 | Straight line graphs

Calculate the gradient and length of a straight line from the coordinates of two points on it; find the equation of a straight line graph in the form $y = mx + c$

Note that this topic becomes a Core topic for the first examination in 2006.

Calculating the gradient of a straight line from its coordinates

In the previous unit we looked at calculating the gradient of a straight line graph by inspection of the graph itself and using the formula. However, it is just as easy to find if we have two sets of coordinates on the line.

Example 1 Find the gradient of a straight line AB which has coordinates A(4, 2) and B(7, 14).

From the given coordinates of A and B we can calculate the corresponding changes in x and y:

$$\text{Change in } x = 7 - 4$$
$$= 3$$
$$\text{Change in } y = 14 - 2$$
$$= 12$$

Use the formula $\text{Gradient} = \dfrac{\text{change in } y}{\text{change in } x} = \dfrac{12}{3} = 4$

This can be illustrated using a sketch.

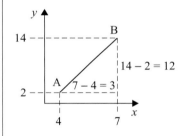

Similarly we can use the formula

$$\text{Gradient} = \frac{y_2 - y_1}{x_2 - x_1}$$

where

x_1 is the first x-coordinate (A)
x_2 is the second x-coordinate (B)
y_1 is the first y-coordinate (A)
y_2 is the second y-coordinate (B)

■ Note that this line has a positive gradient.

Calculating the length of a straight line segment from the coordinates

We can make use of Pythagoras' theorem to calculate the length of a straight line by simply making a rough sketch of it. First let us remind ourselves of the theorem.

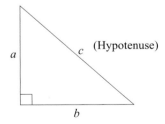

$$c^2 = a^2 + b^2$$
or
$$c = \sqrt{a^2 + b^2}$$

where c is the longest side (hypotenuse) and a, b are the two shorter sides.

Now look at the following example and how we use this formula.

| Example 2 | Find the length of line CD which has coordinates C(2, 1) and D(−6, −5). |

Make a sketch of the line.

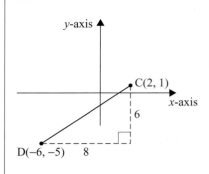

Use the line itself as the longest side (hypotenuse) of the triangle, then the changes in x and y become the lengths of the shorter sides.

Use the formula:

$$\text{Length} = \sqrt{a^2 + b^2}$$
$$= \sqrt{8^2 + 6^2}$$
$$= 10 \text{ units}$$

Therefore length of the line CD = 10 units.

Finding the equation of a straight line

Any linear equation can be written in the form $y = mx + c$. We can break down this equation into its separate parts as shown:

$$y \qquad = \qquad mx \qquad + \qquad c$$

y-value   gradient x-value   intercept with y-axis

We've already looked at how we calculate the gradient either by graph observation or by using two sets of coordinates.

To find the c-value we need to look at the graph and read off the point at which it crosses the y-axis. Look at the example below.

| Example 3 | Find the equation of the following straight line. |

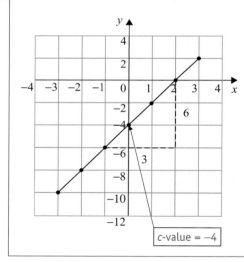

First we find the gradient.

Using the triangle method:

$$\text{Gradient} = \frac{\text{change in } y}{\text{change in } x} = \frac{6}{3}$$
$$= 2$$

Then we find the point where the graph crosses the y-axis.

In this case the c-value = −4.

Therefore the equation of the graph is:

$$y = 2x + (-4)$$
$$y = 2x - 4$$

Straight lines with negative gradients

In Unit 18 we noted that lines with a negative gradient will decrease in y-values as the x-value increases, and vice versa. The method of calculating the gradient is the same.

Example 4 Calculate the gradient of the line below.

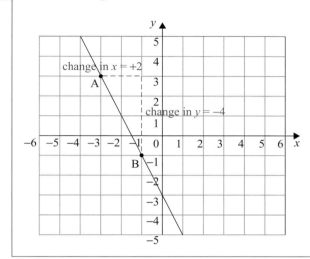

Take any two points on the line and label them A and B. To get from A to B, we need to move $+2$ in the x direction and -4 in the y direction.

$$\text{Gradient} = \frac{\text{change in } y}{\text{change in } x}$$

$$= \frac{-4}{2}$$

$$= -2$$

Finding the equation of a straight line from its coordinates

Using the formula for the equation of a straight line graph ($y = mx + c$) and algebraic manipulation, it is possible to find the equation of a straight line from its coordinates only.

Example 5 A line AB passes through the points A and B with coordinates A $(-5, 4)$ and B $(-3, -2)$. Find the equation of this line.

Using the equation of a straight line graph, $y = mx + c$, where m is the gradient and c is the intercept with the y-axis (the point where the line crosses the y-axis):

$$m = \frac{\text{change in } y}{\text{change in } x} = \frac{(-2) - (4)}{(-3) - (-5)} = \frac{-6}{2} = -3$$

As we do not have a diagram we cannot just write down the value of c. We can calculate it by using either of the coordinate points.

Using point A, where $x = -5$ when $y = 4$ in $y = -3x + c$:

$$(4) = (-3)(-5) + c$$
$$c = -11$$

So the equation of the line is $y = -3x - 11$

20 | Algebraic representation and formulae

Use letters to represent numbers and express basic arithmetic processes algebraically; substitute numbers for letters; simplify expressions; change the subject of a formula

In algebra we use letters and symbols to represent numbers. We can use algebra to obtain or display a general formula and hence use this formula to solve problems. There are certain rules we must follow in algebra.

First let's look at general algebra notation:

$x + y$ A certain quantity x is added to a different quantity y.

$3x$ 3 multiplied by x, e.g. if $x = 2$, $3x = 3 \times (2) = 6$

$4a - 3b$ 4 multiplied by a minus 3 multiplied by b

y^2 y to the power of two or $y \times y$, e.g. if $y = 3$, $y^2 = (3)^2 = 9$

$3c^2 + 4d^3$ $3 \times c \times c + 4 \times d \times d \times d$ (3 multiplied by c squared added to 4 multiplied by d cubed).

Whenever possible, algebra should be written in its shortest possible terms – it should be simplified. The process of simplifying algebraic expressions is called **collecting like terms**.

Collecting like terms (simplification)

When adding or subtracting algebraic expressions, you can only **simplify** or combine terms that have the same letter and power of that letter. The final simplified answers should be in alphabetical order with the highest power first (if applicable):

$x + 2x = 3x$ Letter x has a certain value. If we have one of x added to two xs, we get three xs altogether.

$3y - 2y = y$ This time we are subtracting two ys from three ys.

$3x + 2y + 2x + 4y = 5x + 6y$ The x terms can be added to each other and the y terms can be simplified, but we cannot combine them because they are different letters: x represents one quantity and y represents a different quantity.

Collect like terms

$y^2 + 2y^2 + 3y^3 = 3y^3 + 3y^2$ We can add the first two terms as the letter and power are the same, but the third term has a higher power so it cannot be added to the others.

■ Note that the simplified answers have been arranged in alphabetical order with the highest powers in order from left to right.

Multiplying and dividing algebraic terms

When multiplying terms, first multiply the numbers and then the letters together:

$3a \times 4b = 3 \times 4 \times a \times b = 12ab$

$5c^2 \times 3c^4 = 5 \times 3 \times c \times c \times c \times c \times c \times c = 15c^6$

$2c^2 \times 3d^4 \times 2e^2 = 2 \times 3 \times 2 \times c \times c \times d \times d \times d \times d \times e \times e$
$$= 12c^2d^4e^2$$

When dividing terms, change the terms to their unsimplified form and then cancel where possible:

$$6a^4 \div 2a^2 \quad = \frac{\cancel{6}^3 \times a \times a \times \cancel{a} \times \cancel{a}}{\cancel{2} \times \cancel{a} \times \cancel{a}} = 3 \times a \times a = 3a^2$$

$10x^5 \div 5x^4 \quad = 2x^1 = 2x$

$4x^3y^2 \div x^2y = 4xy$

A quicker method would be to divide the numbers in your head and use indices rules for the letters, e.g. $x^5 \div x^2 = x^{5-2} = x^3$.

■ Revise the rules of indices in Unit 23 to speed up the process of simplification with powers.

Forming simple algebraic statements

We can choose any letter we like to represent the quantities, but x is used most frequently in algebra.

- the sum of two numbers

  $x + y$ x is the first number and y is the second number.

- seven times a number x

  $7x$

- half of a number x

  $\dfrac{x}{2}$

- the sum of two numbers x and y divided by a third number z.

  $\dfrac{x + y}{z}$

- the total cost, x, of two apples each costing a pence and three bananas each costing b pence

  $x = 2a + 3b$

Substitution in algebra

With substitution, we are told the value of the letters and hence we can calculate the actual value of an expression.

| Example 1 | If $x = 2$, $y = 3$ and $z = 4$, find the value of the expressions: (a) $3x$ (b) $2y^2$ (c) $2x - z$ |

(a) $3x = 3 \times (2) = 6$ $3x$ means three multiplied by the value of x.

(b) $2y^2 = 2 \times (3)^2 = 2 \times 9 = 18$

(c) $2x - z = 2 \times (2) - (4) = 4 - 4 = 0$

■ Note that the value of the letters can be negative and/or a fraction value. The process of changing the letters to their number value is still the same.

Transforming simple formulae

There are many mathematical formulae in current use, but what if the formula gives one variable and we need another variable somewhere else in the same formula? In these cases we have to rearrange the formula for our required term. Take, for example, the formula used to find the area of a rectangle:

Width (?)   Area = 20 cm²   Length = 5 cm

Area   = Length × Width

↑ Subject of formula   ↑ Required term

The formula is used to calculate area when we know the values of length and width. We say that area is the **subject** of this formula. For the rectangle above we know the area and the length but not the width, so we have to rearrange the formula to make the width the subject.

To change the subject of the formula we leave the required term where it is and step by step move terms away from it so that it is isolated on one side and becomes the new subject of the formula.

This is the same as solving an equation for a required unknown.

Example 2	Manipulate the formula $x = cy - d$ so that y is the subject.

$$x = cy - d \qquad \text{We need to eliminate the } -d \text{ first.}$$
$$x + d = cy - d + d \qquad \text{Add } d \text{ to both sides.}$$
$$x + d = cy \qquad \text{There are no } d\text{s left on the right-hand side.}$$
$$\frac{x + d}{c} = \frac{cy}{c} \qquad \text{Divide both sides by } c.$$
$$\frac{x + d}{c} = y \qquad y \text{ is now the subject.}$$

When we move a term from one side to the other, we are effectively inverting its operation:

$$+ \longrightarrow - \qquad - \longrightarrow +$$
$$\times \longrightarrow \div \qquad \div \longrightarrow \times$$
$$x^2 \longrightarrow \sqrt{x} \qquad \sqrt{x} \longrightarrow x^2$$

Changing sides $\longrightarrow$ Changing operation

Now we will solve the rectangle problem on page 63.

Area = Length × Width

$$A = LW$$

L changes sides

$$\frac{A}{L} = W$$

We can simplify this further by using the first letter for each term.

To isolate the W we need to move the L to the other side, remembering to use the inverse operation of multiply by L, which is divide by L.

Now W is the subject of the formula.

So $W = \dfrac{20}{5} = 4 \, \text{cm}$

Substitute $A = 20$ and $L = 5$, to give $W = 4 \, \text{cm}$.

The width of the rectangle = 4 cm.

Example 3	For each of the following formulae, rearrange to make the letter in **bold** the subject:

(a) $x = \dfrac{b}{\textbf{\textit{y}}}$ (b) $S = \dfrac{t\textbf{\textit{a}}}{p}$ (c) $E = m\textbf{\textit{v}}^2$

(a) $x = \dfrac{b}{y}$

$xy = b$ Move y to the other side to remove the fraction.

$y = \dfrac{b}{x}$ Isolate y by moving the x to the other side, remembering to use the inverse of multiply by x, which is divide by x.

(b) $S = \dfrac{ta}{p}$

$Sp = ta$ Eliminate the fraction by moving the p to the other side. The inverse operation is multiply by p.

$\dfrac{Sp}{t} = a$ Isolate a by moving the multiply by t to the other side so it becomes divide by t.

(c) $E = mv^2$

$\dfrac{E}{m} = v^2$ Move the multiply by m to the other side to become divide by m.

$\sqrt{\dfrac{E}{m}} = v$ Remove the square by moving it to the other side where it becomes a square root.

The whole process becomes much quicker with practice. Make sure that you use the correct inverse when you move a term from one side of the equation to the other.

The method of rearranging formulae is called **transposition** and is extremely powerful, as we will see when we look at solving equations in Unit 24. It is used in many other topics throughout the syllabus.

20 | Algebraic representation and formulae

Construct equations; transform more complicated formulae

Constructing equations

In order to construct an equation, follow the three simple rules below:

- Represent the quantity to be found by a **symbol**. (x is usually used.)
- Form the **equation** which fits the given information.
- Make sure that both sides of the equation are in the same **units**.

Example 1   Express algebraically: five times a number x minus three times a number y.

$5x - 3y$ Five times x is $5x$ and three times y is $3y$.

Example 2   A girl is m years old now. How old was she 3 years ago?

$m - 3$ Three years before her present age will be m minus 3 years.

Example 3   I think of a number. If I subtract 9 from it and multiply the answer by 4, the result is 32. What is the number I thought of?

$$4(x - 9) = 32$$
$$4x - 36 = 32$$
$$4x = 32 + 36$$
$$x = 17$$

Let the unknown number be x. If I subtract nine from it, it becomes $x - 9$. I then multiply it by four, so $x - 9$ goes inside brackets because the whole expression is multiplied by four. The thirty-two goes on the right-hand side of the equation. Then solve as a linear equation.

Example 4   The sides of a triangle are x cm, $(x - 5)$ cm and $(x + 3)$ cm. If the perimeter is 25 cm, find the lengths of the three sides.

$$x + (x - 5) + (x + 3) = 25 \text{ cm}$$
$$3x - 2 = 25 \text{ cm}$$
$$x = 9 \text{ cm} \quad \text{(one side)}$$
$$(x - 5) = 4 \text{ cm} \quad \text{(second side)}$$
$$(x + 3) = 12 \text{ cm} \quad \text{(third side)}$$

Perimeter of the triangle is the sum of all three sides, so we simply add the three lengths. When we have the first dimension, using substitution, we can find the remaining two sides.

■ Note that we have gone one step further with examples 3 and 4 and solved for x. Solving equations will be looked at in greater detail in Unit 24.

Transforming more complicated formulae

Although the method of 'change sides, change operation' is primarily used again, the formulae are more complicated in the extended syllabus so we need to transform the formulae in the following sequence:

1 Remove square roots or other roots.
2 Remove fractions.
3 Clear brackets.
4 Collect together the terms containing the required subject.
5 Factorise if necessary.
6 Isolate the required subject.

It is unlikely that a formula will contain all six operations, in which case go to the next applicable operation in the list.

Example 5

$M = 5(x + y)$	Make y the subject.
$M = 5x + 5y$	Expand the brackets.
$M - 5x = 5y$	Move the $5x$ to the other side to become $-5x$
$\dfrac{M - 5x}{5} = y$	Move multiply by 5 to the other side to become divide by 5.
$\dfrac{M}{5} - x = y$	Simplify.

Example 6

$y = \dfrac{7}{4 + x}$	Make x the subject.
$(4 + x)y = 7$ $4y + xy = 7$	Remove fractions: move divide by $(4 + x)$ to the other side to become multiply by $(4 + x)$, then expand the brackets.
$xy = 7 - 4y$	Move $+4y$ to the other side to become $-4y$.
$x = \dfrac{7 - 4y}{y}$	Move multiply by y to the other side to become divide by y.
$x = \dfrac{7}{y} - 4$	Simplify.

Example 7

$y = \dfrac{2 - 5x}{2 + 3x}$	Make x the subject.
$(2 + 3x)y = 2 - 5x$	Remove the fraction by taking $(2 + 3x)$ to the other side, and expand the brackets.
$2y + 3xy = 2 - 5x$ $5x + 3xy = 2 - 2y$	Collect the terms involving x on one side; move other terms to the other side.
$x(5 + 3y) = 2 - 2y$	Factorise for x.
$x = \dfrac{2 - 2y}{5 + 3y}$	Move $(5 + 3y)$ to the other side.

■ Note that the correct sequence must be followed: if an operation doesn't apply, then move to the next in sequence.

21 | Algebraic manipulation

Calculate algebra with directed numbers; expand brackets and factorise

In Unit 3 we learnt that directed numbers are numbers with either a positive or a negative sign. When using these numbers in algebra, it is important that we follow the same rules as we would for normal directed number calculations:

(positive number) $\times$ / $\div$ (positive number) = (positive answer)
(negative number) $\times$ / $\div$ (negative number) = (positive answer)
(positive number) $\times$ / $\div$ (negative number) = (negative answer)
(negative number) $\times$ / $\div$ (positive answer) = (negative answer)

So $3y \times (-2y) = -6y^2$ and $-3y \times -2y = 6y^2$.

Expanding brackets

When removing brackets, every term inside the bracket must be multiplied by whatever is outside the bracket.

Example 1	Expand the brackets and simplify where possible:
	(a) $3(x+2)$ (b) $2a(3a+4b-3c)$ (c) $3(x-4)+2(4-x)$

(a) $3(x+2) = 3x + 6$ The expression means three lots of x and three lots of two.

(b) $2a(3a+4b-3c) = 6a^2 + 8ab - 6ac$ $2a$ must be multiplied by every term inside the bracket.

(c) $3(x-4) + 2(4-x) = 3x - 12 + 8 - 2x$ Expand the first bracket and then the second.
$$= x - 4$$
Collect together like terms and simplify.

Example 2	Simplify the following expression:

$$2(2a+2b+2c) - (a+b+c) - 3(a+b+c)$$

$= 4a + 4b + 4c - a - b - c - 3a - 3b - 3c$ Expand each bracket in turn to give all nine terms.

$= 4a - a - 3a + 4b - b - 3b + 4c - c - 3c$ Collect together the like terms to form one
$= 0$ simplified expression.

Factorisation

Factorisation is the reverse of expanding brackets. We start with a simplified expression and put it back into brackets. When factorising, the largest possible factor (number or letter) is removed from each of the terms and placed outside the bracket.

<table>
<tr>
<td>Example 3</td>
<td colspan="2">Factorise each of the following expressions:
(a) $6x + 15$ (b) $10a + 15b - 5c$ (c) $8x^2y - 4xy^2$</td>
</tr>
<tr>
<td></td>
<td>(a) $6x + 15 = 3\left(\dfrac{6x}{3} + \dfrac{15}{3}\right)$</td>
<td>3 is the highest number that will divide into both 6 and 15.</td>
</tr>
<tr>
<td></td>
<td>$= 3(2x + 5)$</td>
<td>Highest number factor is 3. There is no common letter, so only 3 goes outside the bracket.</td>
</tr>
<tr>
<td></td>
<td>(b) $10a + 15b - 5c = 5\left(\dfrac{10a}{5} + \dfrac{15b}{5} - \dfrac{5c}{5}\right)$</td>
<td>5 is the highest number that will divide into 10, 15 and -5.</td>
</tr>
<tr>
<td></td>
<td>$= 5(2a + 3b - c)$</td>
<td>Highest number factor is 5. There is no common letter, so only 5 goes outside the bracket.</td>
</tr>
<tr>
<td></td>
<td>(c) $8x^2y - 4xy^2 \quad = 4xy\left(\dfrac{8x^2y}{4xy} - \dfrac{4xy^2}{4xy}\right)$</td>
<td>$4xy$ is the highest common factor.</td>
</tr>
<tr>
<td></td>
<td>$= 4xy(2x - y)$</td>
<td></td>
</tr>
</table>

<table>
<tr>
<td>Example 4</td>
<td colspan="2">Factorise $4r^3 - 6r^2 + 8r^2s$.</td>
</tr>
<tr>
<td></td>
<td>$= 2r^2\left(\dfrac{4r^3}{2r^2} - \dfrac{6r^2}{2r^2} + \dfrac{8r^2s}{2r^2}\right)$</td>
<td>Highest common factor is $2r^2$.</td>
</tr>
<tr>
<td></td>
<td>$= 2r^2(2r - 3 + 4s)$</td>
<td></td>
</tr>
</table>

21 | Algebraic manipulation

Manipulate harder algebraic expressions, algebraic fractions; factorise using difference of two squares, quadratic, grouping methods

Expanding products of algebraic expressions

These expressions are more complicated, in particular they use double brackets. When expanding double brackets, all terms in the first bracket must be multiplied by all terms in the second bracket.

Example 1 Expand the following and simplify your answer:

(a) $(x + 3)(x + 2)$ (b) $(2y + 1)(2y - 2)$ (c) $(4x + 4y)^2$

(a) $(x + 3)$ $(x + 2)$ $= x^2 + 2x + 3x + 6$ All terms in one bracket must be
   ↑   ↑   ↑   ↑ $= x^2 + 5x + 6$ multiplied by each term in the other
  1st term 2nd term 1st term 2nd term   bracket:
  (B1)   (B1)   (B2)   (B2)
    (Bracket 1)   (Bracket 2)

$$1st\ (B1) \times 1st\ (B2) = x^2$$
$$1st\ (B1) \times 2nd\ (B2) = 2x$$
$$2nd\ (B1) \times 1st\ (B2) = 3x$$
$$2nd\ (B1) \times 2nd\ (B2) = 6$$

(b) $(2y + 1)(2y - 2) = 4y^2 - 4y + 2y - 2$ Same rules as above, simplifying
$$= 4y^2 - 2y - 2$$ the answer where possible.

(c) $(4x + 4y)^2 = (4x + 4y)(4x + 4y)$ The whole of the bracket is squared,
$$= 16x^2 + 16xy + 16xy + 16y^2$$ so we have to write out the bracket
$$= 16x^2 + 16y^2 + 32xy$$ twice and multiply out each term.

Harder factorisation

There are three basic methods:

- factorisation by grouping   Usually involves 4 terms.
- difference of two squares   Involves 2 terms.
- factorisation of quadratic equations. Involves 3 terms.

Examples 2–4 demonstrate each method.

Factorisation by grouping

Example 2 Factorise the following expressions:

(a) $ax + bx + ay + by$ (b) $6x + xy + 6z + zy$ (c) $2x^2 - 3x + 2xy - 3y$

(a) $ax + bx + ay + by = a(x + y) + b(x + y)$ Since $(x + y)$ is a factor of both
$$= (a + b)(x + y)$$ terms we can place a + b inside the
  other bracket.

(b) $6x + xy + 6z + yz = 6(x + z) + y(x + z)$ Now $(x + z)$ is a factor of both
$$= (6 + y)(x + z)$$ terms, so we can place $6 + y$ inside
  the other bracket.

(c) $2x^2 - 3x + 2xy - 3y = 2x(x + y) - 3(x + y)$ $(x + y)$ is a common factor, so this
$$= (2x - 3)(x + y)$$ goes in one bracket and $2x - 3$
  goes in the other.

Difference of two squares

If we expand the expression $(x + y)(x - y)$:

$$(x + y)(x - y) = x^2 - xy + xy - y^2 = x^2 - y^2$$

x^2 and y^2 are both squared terms, so $x^2 - y^2$ is known as the **difference of two squares.**

Example 3 Factorise the following:
(a) $4a^2 - 9b^2$ (b) $144 - y^2$ (c) $16x^4 - 81y^4$

(a) $\begin{aligned} 4a^2 - 9b^2 &= (2a)^2 - (3b)^2 \\ &= (2a + 3b)(2a - 3b) \end{aligned}$ In this case the $2a$ term will represent the 'x' and the $3b$ term will represent the 'y'.

(b) $\begin{aligned} 144 - y^2 &= (12)^2 - y^2 \\ &= (12 + y)(12 - y) \end{aligned}$ Now the 12^2 is the first squared term and the y^2 is the second squared term.

(c) $\begin{aligned} 16x^4 - 81y^4 &= (4x^2)^2 - (9y^2)^2 \\ &= (4x^2 + 9y^2)(4x^2 - 9y^2) \end{aligned}$ The $(4x^2)^2$ term represents the first squared term and the $(9y^2)^2$ term represents the second squared term.

■ Note that each of the above answers can be checked by reversing the process, i.e. by expanding the brackets back to get the original stated question. The last equation becomes:

$$\begin{aligned} (4x^2 + 9y^2)(4x^2 - 9y^2) &= 16x^4 - 36x^2y^2 + 36x^2y^2 - 81y^4 \\ &= 16x^4 - 81y^4 \end{aligned}$$ This is the original problem.

Factorisation of quadratic equations

A quadratic equation is any equation which contains a **squared term**, such as $x^2 + 5x + 6$. Any quadratic equation can be written as a product of two brackets.

Example 4 Factorise the following quadratic equations:
(a) $x^2 + 7x + 12$ (b) $x^2 - 2x - 63$ (c) $3x^2 + 8x + 4$

(a) $x^2 + 7x + 12$

We know that first term in each bracket must be x, since x multiplied by $x = x^2$. In order to find the other terms, follow the method below.

$$x^2 \qquad \underset{\uparrow}{+7x} \qquad \underset{\uparrow}{+12}$$

Sum of the   Product of the
two numbers   two numbers

Products which make 12 are 4×3, -4×-3, 2×6, -2×-6, 1×12, -1×-12.

Sums of two numbers which make 7 are $4 + 3$, $6 + 1$, $5 + 2$.

As 4 and 3 are the only numbers that give the correct sum and the correct product, these must be the missing terms.

$$x^2 + 7x + 12 = (x + 4)(x + 3)$$

■ Note that the order of the numbers is not important.

(b) $$x^2 \qquad \underset{\uparrow}{-2x} \qquad \underset{\uparrow}{-63}$$

Sum   Product

Products which make -63 are 9 and -7, 7 and -9, 3 and -21, -21 and 3, -1 and 63, -63 and 1.

Of these, the sum of two numbers which make -2 are 7 and -9.

As 7 and -9 give both the correct sum and the correct product, these must be the missing terms.

$$x^2 + 2x - 63 = (x + 7)(x - 9)$$

(c) $3x^2$ $+8x$ $+4$ This type of question is more difficult because we only

           ↑   ↑   know that the product is 4 and that the first term in

    Product of   Product of   each bracket is x and $3x$, so we have to list the products

    x and $3x$ two numbers   of 4 and see which two numbers go in the brackets to

                                              give us the middle term of $8x$.

Products which make 4 are 2×2, -2×-2, 1×4, -1×-4.

By trial and error, the only two numbers that will give us $8x$ after expanding the brackets are 2 and 2.

$3x^2 + 8x + 4 = (3x + 2)(x + 2)$

■ Note that you should always check your answers at the end by multiplying out brackets to get back to the original problem.

Manipulating algebraic fractions

The rules for algebraic fractions are the same as for normal fractions.

For example, when multiplying or dividing:

$$\frac{a}{c} \times \frac{b}{d} = \frac{ab}{cd} \qquad\qquad \frac{a}{b} \div \frac{c}{d} = \frac{a}{b} \times \frac{d}{c} = \frac{ad}{bc}$$

When adding or subtracting, if the denominators are the same, we just add/subtract the numerators.

$$\frac{a}{5} + \frac{b}{5} = \frac{a + b}{5} \qquad\qquad \frac{x}{2} - \frac{y}{2} = \frac{x - y}{2}$$

However, when the denominators are different we have to manipulate one of the fractions so that the denominators become equal:

$$\frac{2}{y} + \frac{3}{2y} = \frac{4}{2y} + \frac{3}{2y} = \frac{7}{2y}$$

       $\times 2$

First fraction must be multiplied by 2 to make the denominators equal.

However, in some cases one denominator is not a multiple of the other, so then we have to find a **common multiple** of all the denominators and make this the **common denominator**:

$$\frac{x}{4} + \frac{5y}{9} = \frac{9x}{36} + \frac{20y}{36} = \frac{9x + 20y}{36}$$

36 is a common multiple of 4 and 9.

Even with more complex fractions, the method of finding a common denominator is still required:

$$\frac{1}{x - 2} - \frac{2}{x - 3}$$

$$= \frac{1}{x - 2} \times \frac{x - 3}{x - 3} - \frac{2}{x - 3} \times \frac{x - 2}{x - 2}$$

Now a common denominator is the product of $x-2$ and $x-3$. We multiply $\frac{1}{x-2}$ by $\frac{x-3}{x-3}$ (which is equal to 1) and multiply $\frac{2}{x-3}$ by $\frac{x-2}{x-2}$ (which is also equal to 1). Multiplying by 1 in each case means we have not changed the value of the fractions.

The two fractions now have a common denominator and can be combined and then simplified.

$$= \frac{1(x - 3) - 2(x - 2)}{(x - 2)(x - 3)}$$

$$= \frac{x - 3 - 2x + 4}{(x - 2)(x - 3)}$$

$$= \frac{-x + 1}{(x - 2)(x - 3)}$$

Algebraic manipulation   **EXTENDED**   71

Example 5

Simplify the following algebraic fraction: $\dfrac{5m}{4y} - \dfrac{3m}{6y}$

$$\frac{5m}{4y} - \frac{3m}{6y} = \frac{5m(6y)}{(4y)(6y)} - \frac{3m(4y)}{(4y)(6y)}$$

Express the fractions in terms of a common denominator $(4y)(6y)$.

$$= \frac{30my - 12my}{(4y)(6y)}$$

Multiply out the numerators.

$$= \frac{18my}{24y^2}$$

Simplify the fraction to its lowest terms.

$$= \frac{3m}{4y}$$

Example 6

Simplify $\dfrac{2r}{3x} + \dfrac{5r}{4x} - \dfrac{3r}{2x}$

$$\underset{\times 4}{\frac{2r}{3x}} + \underset{\times 3}{\frac{5r}{4x}} - \underset{\times 6}{\frac{3r}{2x}} = \frac{8r}{12x} + \frac{15r}{12x} - \frac{18r}{12x}$$

Common denominator $= 12x$, so we multiply each fraction by the appropriate number and then simplify the numerators.

$$= \frac{5r}{12x}$$

Simplifying algebraic fractions involving quadratics

When you have an expression such as $\dfrac{x^2 - 2x}{x^2 - 5x + 6}$ factorise the numerator and denominator and then cancel where possible.

Example 7

Simplify the following algebraic fraction: $\dfrac{x^2 - 2x}{x^2 - 5x + 6}$

$$\frac{x^2 - 2x}{x^2 - 5x + 6} = \frac{x(x - 2)}{(x - 3)(x - 2)}$$

$x^2 - 2x$ factorises to $x(x - 2)$.
$x^2 - 5x + 6$ factorises to $(x - 3)(x - 2)$.
Cancel the common terms.

$$= \frac{x}{x - 3}$$

Example 8

Simplifying the following: $\dfrac{x^2 + 4x}{x^2 + x - 12}$

$$\frac{x^2 + 4x}{x^2 + x - 12} = \frac{x(x + 4)}{(x + 4)(x - 3)}$$

$x^2 + 4x$ factorises to $x(x + 4)$.
$x^2 + x - 12$ factorises to $(x + 4)(x - 3)$.
Cancel the common terms.

$$= \frac{x}{x - 3}$$

22 | Functions

Find the input and output of a function, inverse and composite functions

An expression such as $f(x) = 3x - 5$, in which the variable is x, is called a **function of x**. Its numerical value depends on the value of x.

The function above can be shown as a diagram.

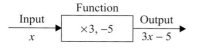

Here the input is a certain value x, the function is a certain operation applied to x and the output is the result after the function has been applied.

There are four basic types of function problems you need to learn.

The type: $f(2) = 3x + 7$

Here the value of x is given (2 in this case) and you have to substitute this value for x in the expression to find the corresponding output.

Example 1	If $f(x) = 2x + 7$, find the value of: (a) $f(3)$ (b) $f(-2)$

(a) $f(3) \quad = 2x + 7$ Use simple substitution of x for the value 3.

$\qquad\qquad = 2(3) + 7 = 13$ In reality, the x input value was 3, we applied the function to it and got the output of 13.

(b) $f(-2) = 2(-2) + 7$

$\qquad\qquad = -4 + 7 = 3$

The type: if $f(x) = 2x + 1$ and $f(x) = 4$, solve for x

Here the value of the output side of the equation is given and you have to find the value of x that was input.

Example 2	If $f(x) = 2x + 1$, solve $f(x) = 4$

We can say $\qquad 4 = 2x + 1$ Solve as a normal linear equation.

$\qquad\qquad\quad 4 - 1 = 2x$ Take the 1 to the other side to become -1, then the multiply by

$\qquad\qquad\quad\quad 3 = 2x$ 2 moves to become divide by 2 and we have the answer.

$\qquad\qquad\quad 1.5 = x$ So the x input must have been 1.5 in order to get an output of 4.

The type $f^{-1}(x)$, the inverse function

This time we have to find the inverse of the function. Below are some functions and their inverses.

Function $f(x)$	Inverse function $f^{-1}(x)$
$x + 3$	$x - 3$
$4x$	$\frac{x}{4}$
$\frac{x+4}{5}$	$5x - 4$

To find the inverse function we can use either a flowchart method or an algebraic method.

The flowchart method

To find the inverse of $f(x) = \frac{x+4}{5}$, first look at the steps needed for x to become $f(x)$:

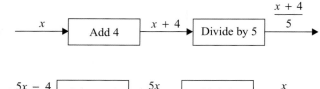

Then, to reverse the function, we perform the reverse operations:

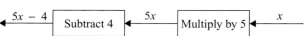

Therefore the inverse $f^{-1}(x) = 5x - 4$

The algebraic method

We can write the function as $f(x) = \frac{x+4}{5}$. Then if we transpose the function to make x the subject we have:

$$f(x) = \frac{x+4}{5}$$

Take the ($\div 5$) to the other side to become ($5f(x)$).

$$5f(x) = x + 4$$

Then the ($+4$) to the other side to become (-4).

$$5f(x) - 4 = x$$

Then simply change the $f(x)$ term back into x.

$$5x - 4 = f^{-1}(x)$$

You now have the inverse function.

The composite functions, $fg(x)$ and $gf(x)$

Here, more than one function is involved.

If $f(x) = 2x - 1$ and $g(x) = 3x + 2$ then we can find $fg(x)$.

As f is the first letter this becomes the main function. We then substitute the $g(x)$ function as the value of x in the $f(x)$ function:

$$fg(x) = 2(3x + 2) - 1$$
$$= 6x + 4 - 1$$
$$= 6x + 3$$

Similarly, we could find the composite function $gf(x)$ by substituting the $f(x)$ function as the value of x in the $g(x)$ function:

$$gf(x) = 3(2x - 1) + 2$$
$$= 6x - 3 + 2$$
$$= 6x - 1$$

These examples use combinations of each type of functions.

Example 3   If $f(x) = 3x - 5$ and $g(x) = 2x + 1$ find:
(a) $f(-2)$ (b) $g(4)$ (c) $g^{-1}(x)$ (d) $fg(3)$

(a) $f(-2) = 3(-2) - 5 = -6 - 5$
$$= -11$$

Both involve simple substitution of x into first the $f(x)$ function, then the $g(x)$ function.

(b) $g(4) = 2(4) + 1 = 8 + 1$
$$= 9$$

(c) $g^{-1}(x) = y = 2x + 1$
$$y - 1 = 2x$$
$$\frac{y-1}{2} = x$$
$$g^{-1}(x) = \frac{x-1}{2}$$

We now find the inverse of the $g(x)$ function using the algebraic method. Try the flowchart method yourself to make sure the answer is correct.

(d) $fg(3) = 3(2x + 1) - 5$
$$= 6x + 3 - 5$$
$$= 6x - 2$$
$$= 6(3) - 2$$
$$= 16$$

The composite function $fg(x)$ is found, where $f(x)$ is the main function; this is $6x - 2$. Then we substitute the value of 3 into the function and get the final answer.

---

23 | Indices

Rules for positive, negative and zero indices

Indices mean **powers** or raised terms. The index figure above a term (number or letter) tells us how many times that term is multiplied by itself. For example:

$4^2 = 4 \times 4$ Number 4 is multiplied by itself.

$y^4 = y \times y \times y \times y$ Letter y is multiplied by itself three times.

Positive, negative and zero indices

Positive indices have positive powers, e.g. x^3 ← positive power

Negative indices have negative powers, e.g. x^{-4} ← negative power

A negative power indicates the reciprocal of the term, e.g. $x^{-4} = \frac{1}{x^4}$

Zero indices have zero powers, e.g. x^0

Any term with a zero index $= 1$, e.g. $x^0 = 1$ It does not matter whether it is a letter or a number.

Rules for indices

- $a^m \times a^n = a^{m+n}$

  e.g. $y^3 \times y^4 = y^{3+4}$ When **multiplying** powers of the same term,
  $= y^7$ we **add** the indices.
  $3^2 \times 3^3 \times 3^4 = 3^{2+3+4}$
  $= 3^9$

- $a^m \div a^n = a^{m-n}$

  e.g. $x^5 \div x^2 = x^{5-2}$ When **dividing** powers of the same term, we
  $= x^3$ **subtract** the indices.
  $4^6 \div 4^3 = 4^{6-3}$
  $= 4^3$

Problems with indices usually involve one or more of the operations above, very often combining positive, negative and zero indices.

Example 1	Simplify the expression $6a^3 \times 4b^{-2} \times 3z^0$.

$6a^3 \times 4b^{-2} \times 3z^0 = 6a^3 \times 4 \times \dfrac{1}{b^2} \times 3 \times 1$ Multiply the numbers together to give 72. Write b^{-2} as $\frac{1}{b^2}$ and z^0 as 1.

$$= \frac{72a^3}{b^2}$$

Exponential equations

These involve terms with unknown indices, e.g. $2^x = 8$, $3^{x+2} = 81$

When solving for the missing index, the terms on both sides of the equation should be made equal.

Example 2	If $2^x = 4$, find the value of x.

First make the terms on each side the same, so change the 4 on the right-hand side to a power of 2.

$$2^x = 2^2$$

As the terms are now the same they can be cancelled and the indices can be equated.

$$\text{So } x = 2$$

Example 3	If $4^x = 64$, find the value of x.

$$4^x = 4^3$$

Make the terms balance, i.e. 4 on both sides, since $4^3 = 64$.

$$\text{So } x = 3$$

Example 4	If $3^{(y+2)} = 27$, find the value of y.

$$3^{(y+2)} = 3^3$$

Change the 27 to a power of 3, i.e. 3^3.

$$\text{So } y + 2 = 3$$

$$y = 1$$

Rearrange the equation for y.

Example 5	If $2^{(-x+3)} = 64$, find the value of x.

$$2^{(-x+3)} = 2^6$$

Change the 64 to a power of 2, i.e. 2^6.

$$\text{So } -x + 3 = 6$$

$$-x = 3$$

$$x = -3$$

Solve for x.

23 | Indices

Fractional indices

This time the indices are in the form of fractions, e.g. $4^{\frac{1}{2}}, 2^{\frac{2}{4}}, 3^{\frac{4}{3}}$.

We need to remember two more rules for fractional indices:

- $(a^m)^n = a^{mn}$

  e.g. $(y^2)^3 = y^{2\times3}$
  $\qquad = y^6$
  $\quad (4^3)^2 = 4^{3\times2}$
  $\qquad = 4^6$

  When raising the power of a quantity to a power, multiply the indices.

- $\sqrt[n]{a^m} = a^{\frac{m}{n}}$

  e.g. $\sqrt[3]{y^2} = y^{\frac{2}{3}}$
  $\quad \sqrt[4]{3^2} = 3^{\frac{2}{4}} = 3^{\frac{1}{2}}$

  To find the nth root of a quantity divide the index of the quantity by n.

Example 1 Evaluate the following without the use of a calculator:

(a) $16^{\frac{1}{4}}$ (b) $81^{\frac{1}{4}}$ (c) $32^{\frac{3}{5}}$ (d) $\dfrac{4^{\frac{5}{6}} \times 4^{\frac{1}{6}}}{4^{\frac{1}{2}}}$

(a) $16^{\frac{1}{4}} = \sqrt[4]{16} = 2$ The fourth root of $16 = 2$, i.e. $2 \times 2 \times 2 \times 2 = 16$.

(b) $81^{\frac{1}{4}} = \sqrt[4]{81} = 3$ The fourth root of $81 = 3$, i.e. $3 \times 3 \times 3 \times 3 = 81$.

(c) $32^{\frac{3}{5}} = 2^{5 \times \frac{3}{5}} = 2^3$ Change 32 to a power of 2, then cancel the indices to leave two
$\qquad = 8$ cubed, which is 8.

(d) $\dfrac{4^{\frac{5}{6}} \times 4^{\frac{1}{6}}}{4^{\frac{1}{2}}} = \dfrac{2^{\frac{2\times5}{6}} \times 2^{\frac{2\times1}{6}}}{2^{\frac{2\times1}{2}}}$ Change the 4s to powers of 2 and multiply each by their index, cancelling where possible.

$\qquad = \dfrac{2^{\frac{5}{3}} \times 2^{\frac{1}{3}}}{2}$ Simplify the top of the fraction by adding the indices (rule for multiplying indices).

$\qquad = \dfrac{2^2}{2}$ Divide the numerator by the denominator to get the final answer.

$\qquad = 2$

■ Note that when we have questions with numbers as the terms, we can check the answers with a scientific calculator, but always show the method.

---

24 | Solution of equations and inequalities

Solve linear equations; solve simultaneous linear equations

Solving linear equations

The method for solving a linear equation is basically the same as for transforming formulae, but we are now given enough information to completely solve it.

Consider the linear equation $x + 4 = 9$

We can rearrange the formula so that x becomes the subject:

$$x = 9 - 4 = 5 \quad \text{which is the solution for } x.$$

This method of 'change sides, change operation' (see Unit 20) is used when solving linear equations, although the equations do get a little more difficult.

Example 1 Solve the following linear equations:

(a) $3x + 4 = 19$ (b) $2(y - 3) = 6$ (c) $6x + 9 = 3x - 54$

(a) $\quad 3x + 4 = 19$	Move the $+4$ to the other side to become -4.
$\qquad 3x = 19 - 4$	
$\qquad 3x = 15$	Take the multiply by 3 to the other side to become divide by 3.
$\qquad x = 5$	
(b) $2(y - 3) = 6$	Expand the brackets, take the -6 to the other side to become $+6$,
$\qquad 2y - 6 = 6$	then take multiply by 2 to the other side to become divide by 2.
$\qquad 2y = 6 + 6$	
$\qquad y = 6$	
(c) $\quad 6x + 9 = 3x - 54$	Take the $+3x$ to the other side to become $-3x$ and the $+9$ to the
$\qquad 6x - 3x = -54 - 9$	other side to become -9, then simplify. Move the multiply by 3 to
$\qquad 3x = -63$	the other side to become divide by 3.
$\qquad x = -21$	

Example 2 Solve the equation for x: $\dfrac{7 + 2x}{3} = \dfrac{9x - 1}{7}$

$\dfrac{21(7 + 2x)}{3} = \dfrac{21(9x - 1)}{7}$	Eliminate fractions by multiplying by the LCM (21).
$7(7 + 2x) = 3(9x - 1)$	
$49 + 14x = 27x - 3$	Expand brackets, move the xs to one side and numbers to the other
$49 + 3 = 27x - 14x$	side, remembering to invert the operations.
$52 = 13x$	
$4 = x$	

Solving simultaneous equations

This time we are looking for the value of two unknowns, usually x and y. There are two methods for solving simultaneous equations. Both are used in this example.

Example 3

Find the values of x and y in the following simultaneous equations:

$$2x + y = 14 \quad \text{(i)}$$
$$x + y = 9 \quad \text{(ii)}$$

1st method (elimination)

With this method we have to eliminate either the x or the y term by adding or subtracting the equations.

In the equations above we can eliminate the y terms by subtracting the second equation from the first:

$$\begin{array}{r} 2x + y = 14 \\ -x + y = 9 \\ \hline x + 0 = 5 \end{array}$$

Therefore $x = 5$. Now we substitute the value for x we have just found into either of the equations:

Using the first equation:
$$2x + y = 14$$
$$2(5) + y = 14$$
$$10 + y = 14$$
$$y = 4$$

So $\quad x = 5$
$\quad\quad y = 4$

2nd method (substitution)

For this method you need to have a good grasp of algebra manipulation.

We have to rearrange one of the equations to make either x or y the subject and then substitute into the other equation.

Choosing the second equation, make x the subject: $\quad x + y = 9$
$$x = 9 - y$$

Now we substitute this expression for x into the first equation:

$$2(9 - y) + y = 14$$
$$18 - 2y + y = 14$$
$$18 - y = 14$$
$$18 - 14 = y$$
$$4 = y$$

Now we have the y-value, we substitute this into the second equation to get the x-value:

$$x + y = 9$$
$$x = 5$$

Example 4

Solve the following simultaneous equations:
$$2x = 18 - 5y \quad \text{(i)}$$
$$x + 3y = 10 \quad \text{(ii)}$$

Using the substitution method, we rearrange the second equation for x:

$x = 10 - 3y \quad$ $3y$ has been moved to the other side to become $-3y$.

$2(10 - 3y) = 18 - 5y \quad$ Now we substitute this expression for x into the other equation to find the value of y.

$20 - 6y + 5y = 18$

$20 - 18 = y \quad$ Once we have the value of y we can substitute it into the second

$2 = y \quad$ equation to find the x-value.

$x + 3(2) = 10 \quad$ Now we check our values: $\quad 2(4) = 18 - 5(2)$

$x = 4 \quad\quad\quad\quad\quad\quad\quad\quad\quad\quad 8 = 8 \quad$ Correct

Solving simultaneous equations which have different coefficients

We will now have to manipulate one or both of the equations so that one pair of coefficients is equal and then use the elimination method as before.

Example 5

Solve for x and y: $4x + y = 14$ (i)
$6x - 3y = 3$ (ii)

$3(4x + y) = 3 \times 14$
$12x + 3y = 42$ (iii)

We can make the coefficients of y equal by multiplying the first equation by 3.

Then using the elimination method:

$12x + 3y = 42$ (iii)
$+6x - 3y = 3$ (ii)
$\overline{18x = 45}$
$x = 2.5$

We add the manipulated equation (iii) to equation (ii). The ys are now eliminated.

Using $x = 2.5$,

$6(2.5) - 3y = 3$
$15 - 3y = 3$
$y = 4$

Substituting the value of x into equation (ii) finally gives us the value of y.
Don't forget to check!

Example 6

Solve for x and y: $5x - 3y = -0.5$ (i)
$3x + 2y = 3.5$ (ii)

$2(5x - 3y) = 2(-0.5)$
$10x - 6y = -1$ (iii)

Multiply the first equation by 2 and the second equation by 3. This gives equal y coefficients.

$3(3x + 2y) = 3 \times 3.5$
$9x + 6y = 10.5$ (iv)

$10x - 6y = -1$
$+9x + 6y = 10.5$
$\overline{19x = 9.5}$
$x = 0.5$

By adding equations (iii) and (iv), the ys are eliminated and x can be found.

$5(0.5) - 3y = -0.5$
$2.5 - 3y = -0.5$
$y = 1$

Substituting our known value of x into the first equation finally gives us the value of y.

■ Note that we could have eliminated the x coefficients first by making them both equal to 15 (equation (i) multiplied by 3 and equation (ii) multiplied by 5). We would have obtained the same values for x and y.

24 Solution of equations and inequalities

Solve quadratic equations by factorisation and *either* by use of the formula *or* by completing the square; solve simple linear inequalities

Solution of quadratic equations

There are three methods of solving quadratic equations:

- factorisation
- using the formula
- completing the square

We mentioned in Unit 21 (Extended) that a quadratic equation takes the form $ax^2 + bx + c$ where a, b and c are constants. We also looked at the factorising method, so this will not be examined again in detail. But we need to look at how we actually solve a quadratic equation once we have it in brackets form.

Example 1 Solve the equation $x^2 + 3x - 10 = 0$.

The squared term x^2 tells us that the equation has two possible values. First we factorise:

$$x^2 + 3x - 10 = 0 \longrightarrow (x + 5)(x - 2) = 0 \qquad \text{Factorising method.}$$

Therefore, if $(x + 5)(x - 2) = 0$ either $(x + 5) = 0$ or $(x - 2) = 0$

So $x = -5$ or $x = 2$

Now perform a check.

If $x = -5$,   If $x = 2$
$(-5)^2 + 3(-5) - 10 = 0$ $(2)^2 + 3(2) - 10 = 0$ Since the left-hand side of the equation is
$\qquad 25 - 15 - 10 = 0$ $\qquad 4 + 6 - 10 = 0$ equal to the right-hand side we know the
$\qquad\qquad\quad 0 = 0$ $\qquad\qquad\quad 0 = 0$ answers obtained for x are correct.

We can solve quadratic equations either by using the completing the square method or by using the quadratic formula. We look at the quadratic formula method on page 81.

The quadratic formula

For a quadratic of the form $ax^2 + bx + c$:

$$x = \frac{-b \pm \sqrt{b^2 - 4ac}}{2a} \qquad \text{where } a, b \text{ and } c \text{ are the constant values}$$

Example 2   Solve the equation $x^2 + 5x - 7 = 0$, giving your answers to 2 decimal places.

Comparing this to the standard quadratic equation $ax^2 + bx + c$, we see that $a = 1, b = 5, c = -7$.

Using the formula $x = \dfrac{-b \pm \sqrt{b^2 - 4ac}}{2a}$ we can now substitute the values of a, b and c into it:

$$= \frac{-(5) \pm \sqrt{(5)^2 - 4 \times (1) \times (-7)}}{2 \times 1}$$

$$= \frac{-5 + \sqrt{25 + 28}}{2} \qquad \text{or} \qquad \frac{-5 - \sqrt{25 + 28}}{2} \qquad$$ Again we have two possible values for x. Both need to be checked.

$$= \frac{2.28}{2} \qquad \text{or} \qquad \frac{-12.28}{2}$$

$$= 1.14 \qquad \text{or} \qquad -6.14$$

If $x = 1.14$
$(1.14)^2 + 5(1.14) - 7 = 0$
$1.2996 + 5.7 - 7 = 0$
$6.9996 - 7 = \text{approx. } 0$

■ Note that the left-hand side doesn't exactly equal zero due to rounding up errors in our calculator but it is very close.

If we do not have the a or b constant values, we would substitute zero for these terms in the formula.

Solving linear inequalities

We looked at simple inequalities in Unit 5 and will see graphical inequalities in Unit 25, so make sure that you are familiar with the symbols.

When solving an inequality treat it as a normal equation, only considering the symbol at the last stage of your working.

Example 3   Solve the inequalities   (a) $x + 3 < 7$ (b) $8 \leq x + 1$.

(a) $x + 3 < 7$ Move $+3$ to other side to become -3, simplify to obtain the solution *and*
    $x < 7 - 3$ other possible solutions for x that still satisfy the inequality.
    $x < 4$

So, x could be 3, 2, 1, . . .

(b) $8 \leq x + 1$ Move $+1$ to other side to become -1, simplify, then solve for all possible
    $8 - 1 \leq x$ values of x.
    $7 \leq x$

So, x could be 7, 8, 9, . . .

Example 4   Solve the inequality $9 - 4x \geq 17$.

$9 - 4x \geq 17$ The $+9$ is moved to the other side where it becomes -9, the multiply by -4
$-4x \geq 17 - 9$ goes to the other side where it becomes divide by -4.
$-4x \geq 8$
$x \leq -2$ **Invert the inequality to eliminate the minus sign.**

So, x could be $-2, -3, -4, . . .$ Possible solutions for x.

Solving harder inequalities

Take, for example, the inequality $5 < 3x + 2 \leq 17$. Here x has a range of values that satisfy the inequalities $3x + 2 > 5$ and $3x + 2 \leq 17$.

Example 5	Find the range of values for the equation $5 < 3x + 2 \leq 17$.

1st part of inequality	**2nd part of inequality**

$5 < 3x + 2$	Move +2 to the other side	$3x + 2 \leq 17$	Move +2 to the other side
$5 - 2 < 3x$	to become -2, move	$3x \leq 17 - 2$	to become -2, move
$1 < x$	multiply by 3 to the other	$x \leq 5$	multiply by 3 to the other
	side to become divide by		side to become divide by
	3, and solve for x.		3, and solve for x.

Now we can see that x is both greater than 1 and equal to or less than 5, shown as an inequality:

$$1 < x \leq 5$$

So, the x-values are $x = 2, 3, 4, 5$.

The process of solving inequalities is very similar to that of solving linear equations, but remember that when you have a negative x-value, make it positive by moving it to the other side of the equation.

When you have to find a range of values as in Example 5, treat the problem as two separate inequalities, solve both and then use your answers to form one general solution(s).

25 | Linear programming

Represent inequalities graphically and use this representation in the solution of simple linear programming problems (the conventions of using broken lines for strict inequalities and shading unwanted regions will be expected)

Representing inequalities graphically

In Unit 5 we looked at the symbols $>$, $<$, $\geq$, $\leq$.

$a > b = a$ is greater than b $a < b = a$ is less than b

$a \geq b = a$ is greater than or equal to b $a \leq b = a$ is less than or equal to b

Strict inequalities use the signs $>$ and $<$ only.

When drawing the strict inequalities graphs, the actual graph lines are drawn with broken lines. If the inequality uses $\geq$ or $\leq$, then the graphs are drawn with solid lines.

Consider $x \geq 1$, where x could be 1 or any value greater than 1, e.g. 2, 3, 4. We can demonstrate this graphically as:

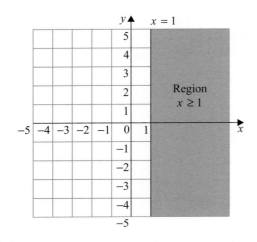

We can see from the graph that the shaded region shows the values of x which are greater than or equal to 1.

The graph of $x = 1$ is a solid line. This is because 1 is included in the inequality.

If we had to show the graph of $x > 1$, the line would be broken because the value of $x = 1$ is not included in the inequality.

■ Note that we can also show an inequality as the unshaded region. The question will tell us whether the region should be unshaded or shaded.

The *unshaded* region shows all the values of $x \leq 2$.

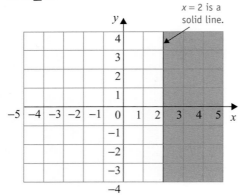

The *unshaded* region shows the values of $y > -1$.

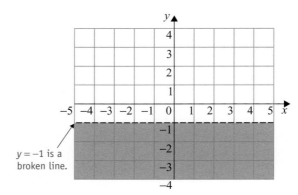

Now we will look at combining inequalities on the same axis.

Example 1 Leave unshaded the regions which satisfy the following inequalities simultaneously:
$x \leq 2, \quad y > -1, \quad y \leq 3, \quad y \leq x + 2.$

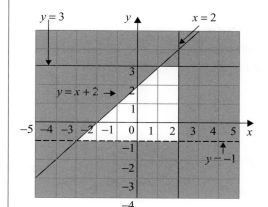

Note how we have drawn the previous graphs on the same axis and then added the last inequality of $y \leq x + 2$.

The unshaded region in the middle obeys all of the inequalities simultaneously.

■ Remember that the way the question is asked is very important – whether you need to identify the shaded or the unshaded region. Always show strict inequalities with broken lines and 'not strict' inequalities with solid lines.

26 | Geometric terms and relationships

Calculate angles; understand and use point, line, parallel and bearing; right, acute, obtuse and reflex angles; perpendiculars, similarity, congruence; types of triangles, quadrilaterals, circles and polygons, and solid figures including nets

Point, line, parallel, bearing

Many of these terms are covered in more depth in Unit 29, but this is a quick summary.

Point

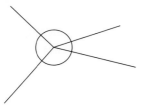

Angles at a point add up to 360°.

Line

When θ is 180°, it's a straight line.

Parallel line angles

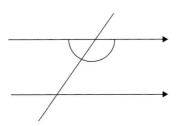

Angles in parallel lines follow certain rules which are examined in detail in Unit 29.

Bearing

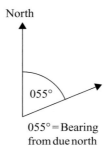

$055° =$ Bearing from due north

A bearing is a *three-figure* angular direction which is always measured from north in a clockwise direction. It can also incorporate a distance by using a suitable scale.

Right angles, acute, obtuse and reflex angles

These are the names of the different types of angles.

Right angles

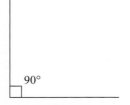

An angle of 90° is a right angle.

Acute angles

An angle between 0° and 90° is an acute angle.

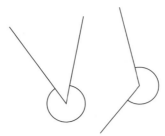

An angle between 90° and 180° is an obtuse angle.

An angle greater than 180° is called a reflex angle.

Perpendicular lines

Perpendicular lines are at 90° to each other, i.e. they cross each other at right angles.

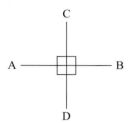

AB is perpendicular to CD.

Similarity and congruence

If shapes are similar, their corresponding angles are equal and the ratio of their corresponding sides is equal.

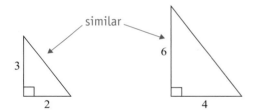

The second triangle has corresponding angles the same as those in the first triangle, but it is exactly twice the size of the first triangle.

Congruent means identical to, i.e. the corresponding angles are equal and the corresponding sides are equal.

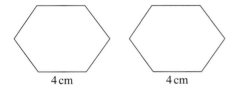

These shapes are congruent hexagons.

Vocabulary of triangles and quadrilaterals

The various triangles (equilateral, isosceles, scalene, right-angled) and the different quadrilaterals (square, rectangle, parallelogram, trapezium, kite and rhombus) are considered in more detail in Unit 29.

Circles and polygons

These are studied in more detail in Unit 29.

Simple solid figures and their nets

These shapes are studied in detail in Unit 29, but now we need to look at some of their nets.

A **net** is an unfolded shape. Imagine that you have a solid shape and you to need to unfold it.

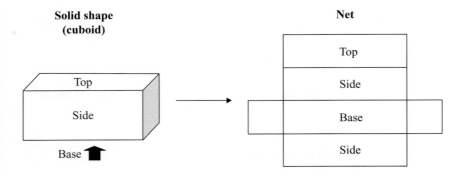

You can investigate this at home with any cuboid-shaped solid, such as a packet of cornflakes.

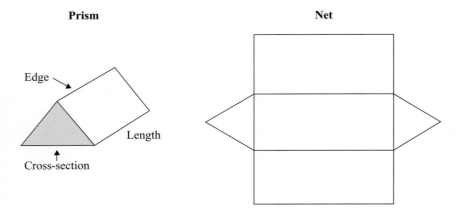

When drawing the net always ensure it is the correct size, i.e. the length dimension corresponds to the length dimension on the net, and so on. It also helps to count the faces to make sure you have accounted for all of them, for example:

- A cube or cuboid has 6 faces.
- A cylinder has 2 circular ends and a rectangular face.
- A triangular prism has 5 faces.

26 | Geometric terms and relationships

Areas and volumes of similar triangles, figures and solids

Similar shapes

Two shapes are similar if:

- all the corresponding angles are equal
- the ratios of corresponding sides are equal

The shapes below are similar.

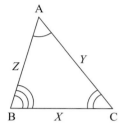

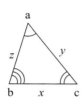

In these shapes the corresponding angles are equal, i.e. $\widehat{A} = \widehat{a}$
$$\widehat{B} = \widehat{b}$$
$$\widehat{C} = \widehat{c}$$

The ratios of corresponding sides are also equal, i.e. if $X = 2x$
then $Y = 2y$
$$Z = 2z$$

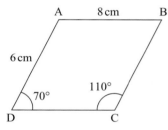

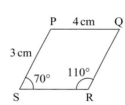

Again the corresponding angles are equal, i.e. $\text{ADC} = 70°$
$$\text{PSR} = 70°$$

Also the ratios of the corresponding sides are equal, i.e. $\text{AB} = 2\text{PQ}$
$$\text{AD} = 2\text{PS}$$

Example 1   Find the length x in these similar shapes.

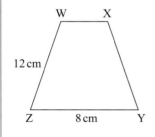

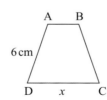

We can see from the shapes that $\text{WZ} = 2\text{AD}$, so we can say that:

$$2\text{DC} = \text{ZY}$$
$$\text{DC} = \text{ZY} \div 2$$
$$= 8\,\text{cm} \div 2$$
$$= 4\,\text{cm}$$

Alternatively,

$$\frac{\text{DC}}{\text{ZY}} = \frac{\text{AD}}{\text{WZ}}$$
$$\text{DC} = \frac{\text{ZY} \times \text{AD}}{\text{WZ}}$$
$$\text{DC} = \frac{8\,\text{cm} \times 6\,\text{cm}}{12\,\text{cm}}$$
$$= 4\,\text{cm}$$

Volumes and surface areas of similar solids

Two solids are similar if they are the same shape and the ratios of their corresponding linear dimensions are equal. Two similar solids are shown below.

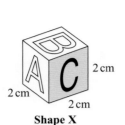 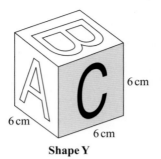

Shape X   **Shape Y**

The linear dimensions of cube Y are three times those of cube X.

If we know that solid shapes, X and Y, are similar, we can apply the following formulae:

$$\frac{\text{Surface area of X}}{\text{Surface area of Y}} = \frac{(\text{Linear dimension of X})^2}{(\text{Linear dimension of Y})^2}$$

$$\frac{\text{Volume of X}}{\text{Volume of Y}} = \frac{(\text{Linear dimension of X})^3}{(\text{Linear dimension of Y})^3}$$

We can calculate a missing area or volume or missing side using the formulae above provided we have the other terms in the formulae.

Example 2   Sphere X has a diameter of 12 cm, a surface area of 450 cm^2 and a volume of 900 cm^3. Sphere Y has a diameter of 6 cm. What is its surface area and volume?

To find the surface area use the formula:

$$\frac{\text{Surface area of Y}}{\text{Surface area of X}} = \frac{(\text{Linear dimension of Y})^2}{(\text{Linear dimension of X})^2}$$

$$\frac{\text{Surface area of Y}}{450} = \frac{(6)^2}{(12)^2}$$

$$\text{Surface area of Y} = \frac{450 \times (6)^2}{(12)^2}$$

$$= 112.5 \, \text{cm}^2$$

To find the volume use the formula:

$$\frac{\text{Volume of Y}}{\text{Volume of X}} = \frac{(\text{Linear dimension of Y})^3}{(\text{Linear dimension of X})^3}$$

$$\text{Volume of Y} = \frac{900 \times (6)^3}{(12)^3} = 112.5 \, \text{cm}^3$$

■ A linear dimension is any length. In this case the diameter could be used, but we could also use the height or base of a triangle, or the length of a cuboid, etc.

Example 3   A circle of 14 cm radius has an area of 616 cm^2. What is the area of a circle with a radius of 7 cm?

$$\frac{\text{Area of circle X}}{\text{Area of circle Y}} = \frac{(\text{Linear dimension of X})^2}{(\text{Linear dimension of Y})^2}$$

$$\text{Area of X} = \frac{616 \times (7)^2}{(14)^2} = 154 \, \text{cm}^2$$

Example 4

A spherical cap has a height of 4 cm and a volume of 64 cm³. A similar cap has a volume of 512 cm³. Find its height.

Rearranging the volume formula to give the linear dimension:

$$\frac{\text{Volume of Y}}{\text{Volume of X}} = \frac{(\text{Linear dimension of Y})^3}{(\text{Linear dimension of X})^3}$$

$$\sqrt[3]{\frac{\text{Volume of Y} \times (\text{Linear dimension of X})^3}{\text{Volume of X}}} = \text{Linear dimension of Y}$$

$$\sqrt[3]{\frac{512 \times (4)^3}{64}} = \text{Linear dimension of Y}$$

$$\text{Height of Y} = 8\,\text{cm}$$

Example 5

The volume of a cone of height 14.2 cm is 210 cm³. Find the height of a similar cone whose volume is 60 cm³.

This problem uses the same rearranged formula as Example 4.

$$\text{Height} = \sqrt[3]{\frac{60 \times (14.2)^3}{210}} = 9.35\,\text{cm}$$

27 | Geometrical constructions

Measure and construct lines, angles, bisectors and simple shapes using ruler, protractor and set of compasses; scale drawings

You should be able to use a ruler to draw and measure lines to a specific length and also to construct and measure angles using a protractor. Take care, however, to read the correct scale on a protractor. This is how to draw an angle of 26°.

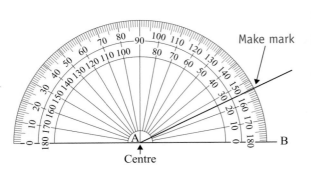

Make mark

Centre

First draw the base line AB (size not important).

Place the centre of the protractor on one end of the line (point A), lining up the zero of the protractor with line AB.

As the zero is on the inside scale, read in an anti-clockwise direction until you reach 26° on the inside scale. At this point, make a mark.

Remove the protractor and draw a straight line from point A through this mark.

Constructing a triangle using ruler and compasses

We'll construct this triangle.

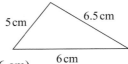

Step 1   Draw the base of the triangle using ruler and pencil (6 cm).

Step 2

    1   Choose any side, e.g. 6.5 cm. Set the compasses to 6.5 cm and draw an arc above the base of the triangle.

    2   Set the compasses to 5 cm and draw an arc above the base from the other end.

    3   Join the point of intersection to the two ends of the base line.

1 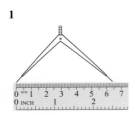   **2** 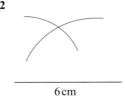   **3**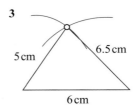

The construction of other simple geometrical figures will be similar. Always choose a base line to start from. If you are given angles to draw, use a protractor unless they are 90° angles, in which case you could use compasses again (see perpendicular bisector section).

Constructing angle bisectors

Bisect means cut in half. First draw the angle using a protractor, then follow these steps.

Step 1   With the compasses set to any value, place the point at the corner of the angle and draw an arc so that it cuts both lines.

Step 2   Keeping the compasses at the same setting, put the point of the compasses on the previous arc cuts and draw two more arcs in the centre of the angle.

Step 3   Draw the line from the corner of the angle to where the arcs cross.

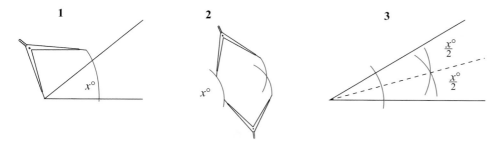

Constructing perpendicular bisectors

A perpendicular bisector is a line which crosses another line at 90°, cutting it in half.

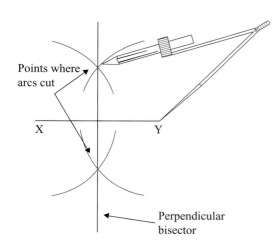

Step 1 Draw the line you are going to bisect (XY).

Step 2 Set the compasses to approximately three-quarters of the length of XY.

Step 3 Place the point of the compasses on one end of the line, e.g. point X. Draw an arc above and below XY.

Step 4 Without changing the compasses' setting, place the point of the compasses on the other end (point Y). Draw an arc above and below XY to cut the previous arcs.

Step 5 Join up the two points where the arcs cross.

Scale drawings

Drawings are made to scale to show accurate representations of larger or smaller images to a manageable size. When you are presented with a scale drawing you are given the scale used and from this you can make accurate calculations. Unit 10 explains how to calculate real lengths from scaled lengths and vice versa. When you have the lengths, it is a simple case of drawing using the techniques covered in this unit.

Drawing squares and rectangles

Only 90° angles and straight lines are used to draw squares and rectangles, which means these can be constructed with a ruler and a set of compasses or even a set square.

Example 1　Use a set square and a ruler to draw a rectangle with edge lengths of 4 cm and 3 cm.

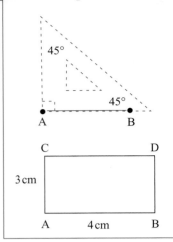

First use a ruler to draw a line AB, 4 cm long.

Now place the edge of the set square along the line AB so that the edge of the rectangle will start at point A as shown in the diagram.

From A draw a line along the edge of the set square. Mark a point C on this line which is 3 cm from A. Repeat at point B to reach a point D. Join C to D with a ruler. Measure the length of CD as a check.

Example 2　Construct the following shape, using a ruler, set square and compasses.

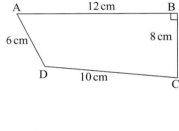

Draw line AB of length 12 cm with the ruler, ensuring enough space is left below it.

Construct the 90° angle at B using the set square and drawing a line of 8 cm to point C.

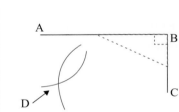

From point C, set the compasses to 10 cm and draw an arc to the left of the shape.

From point A, set the compasses to 6 cm, draw an arc to intersect the previous arc. This will be point D.

Join A to D and D to C. Check the dimensions are correct.

28 | Symmetry

Rotation and line symmetry (including order of rotation symmetry) in two dimensions and symmetry properties of triangles, quadrilaterals and circles

Symmetry means 'reflection', i.e. an image that is identical to the original object.

There are two main types of symmetry:

- line symmetry
- rotation symmetry

Line symmetry

Line symmetry is also called reflection symmetry. Look at this equilateral triangle.

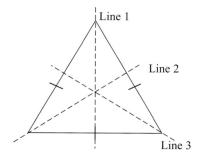

This shape has 3 lines of symmetry: in lines 1, 2 and 3.

Either side of each line, the reflection is identical, in this case a right-angled triangle.

Different shapes have different numbers of lines of symmetry. For example, a circle has an infinite number of lines of symmetry, while a square has 4 lines of symmetry.

Rotation symmetry

Many shapes can be rotated about their centre and still fit their original outline. The rectangle below illustrates this.

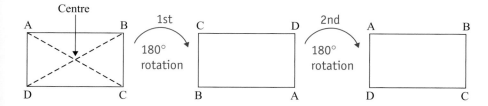

The first rotation is 180° clockwise and the image is identical although the corners have changed position. The second rotation is another 180° clockwise rotation, back to the original position. We can say that the rectangle has **rotation symmetry of order 2**.

Only shapes with **order 2 or greater** are classed as having rotation symmetry. (A 360° turn only is not sufficient.)

Triangles and quadrilaterals

We have already shown that an equilateral triangle has three lines of symmetry and rotation symmetry of order 3. Any regular polygon has line symmetry equal to its rotation symmetry. For example, a regular pentagon has five lines of symmetry and rotation symmetry of order 5 also.

Now we can consider the basic triangles and quadrilaterals and their respective symmetries.

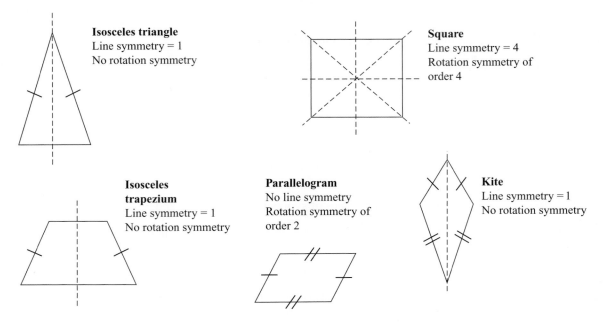

Isosceles triangle
Line symmetry = 1
No rotation symmetry

Square
Line symmetry = 4
Rotation symmetry of order 4

Isosceles trapezium
Line symmetry = 1
No rotation symmetry

Parallelogram
No line symmetry
Rotation symmetry of order 2

Kite
Line symmetry = 1
No rotation symmetry

When you look at the shapes you can quickly see whether or not they are symmetrical. Sometimes, however, it is easier to trace the shape first and then practise folding it over to see if there are lines of symmetry. You can rotate it about its centre to check for rotation symmetry.

Remember these basic rules when describing a shape's symmetry.

- **Line of symmetry** is a mirror line that seems to cut the shape into identical halves when drawn.
- **Rotation symmetry** is the number of times a shape can be rotated about its centre and still look the same.
- All shapes have a rotation symmetry of at least 1, i.e. they can all make a full turn.
- For **regular shapes** (identical sides, identical angles), line symmetry = order of rotation symmetry.

28 | Symmetry

Solid shape symmetries, equal chords, perpendicular/tangent symmetry

Symmetry properties of the prism, cylinder, pyramid and cone

A solid has **plane symmetry** if it can be cut into two halves and each half is the mirror image of the other. The plane separating the two halves is called the **plane of symmetry**.

Here are the basic shapes with their planes of symmetry.

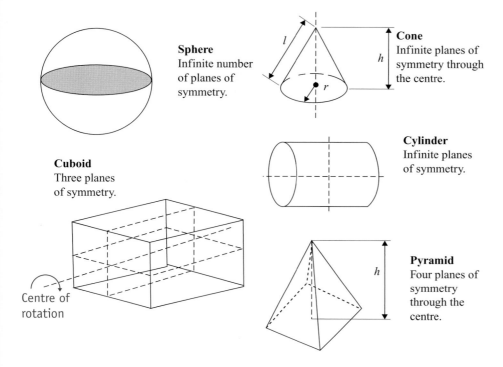

Sphere
Infinite number of planes of symmetry.

Cone
Infinite planes of symmetry through the centre.

Cylinder
Infinite planes of symmetry.

Cuboid
Three planes of symmetry.

Centre of rotation

Pyramid
Four planes of symmetry through the centre.

The solids above also have rotation symmetry because when they are rotated about a central axis, they look the same. For example, the sphere has an infinite order of rotation symmetry, the cuboid has rotation symmetry of order 4 and the pyramid also has rotation symmetry of order 4 (because it has a square base). The cone and the cylinder have an infinite order of rotation symmetry.

Equal chords and perpendicular bisectors

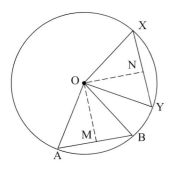

In this diagram of a circle, chords AB and XY are equal. Radii OA, OB, OX and OY are equal, so the triangles OAB and OXY are congruent isosceles triangles. Therefore,

line of symmetry OM = line of symmetry ON

So we can say that OM and ON are perpendicular bisectors of AB and XY respectively.

Tangents to a circle from an external point

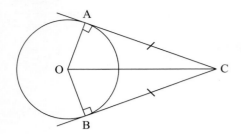

Triangles OAC and OBC are congruent because the angles at A and B are both right angles (OA = OB and OC is common to both triangles). Since OC is common to both triangles,

tangent AC = tangent BC

■ Tangents drawn to the same circle from the same external point are equal in length.

29 | Angle properties

Calculate unknown angles at a point and in parallel lines, shapes and circles

Angles at a point

Angles at a point add up to $360°$. Using this rule we can calculate a missing angle by subtracting the remaining angles from $360°$.

| Example 1 | Find the angle x in this diagram. |

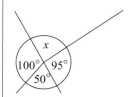

The angles at a point add up to $360°$, so

$$x = 360° - 100° - 95° - 50°$$
$$x = 115°$$

Angles within parallel lines

Angles within parallel lines follow certain rules:

$a + b = 180°$ adjacent angles

$\left.\begin{array}{l} a = d \\ b = c \\ e = h \\ g = f \end{array}\right\}$ vertically opposite angles

$\left.\begin{array}{l} c = f \\ d = e \end{array}\right\}$ alternate angles

$\begin{array}{l} c + e = 180° \\ d + f = 180° \end{array}$ supplementary angles

$\left.\begin{array}{l} a = e \\ b = f \end{array}\right\}$ corresponding angles

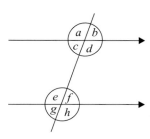

| Example 2 | Find the missing angles in this diagram. |

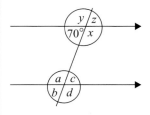

$x = 180° - 70° = 110°$
(angles on a straight line)

$z = 70°$
(z is vertically opposite to $70°$)

$b = 70°$
(a and b are angles on a straight line)

$y = 110°$
(y is vertically opposite to x)

$a = 180° - 70° = 110°$
(a is supplementary to $70°$)

$c = 70°$ and $d = 110°$
(similar reasons to above)

Angle properties of triangles

The triangle has three sides and the sum of its angles is 180°.

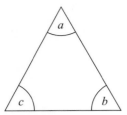

$$a + b + c = 180°$$

The different types of triangle are shown below.

Isosceles triangle

Two equal base angles, two equal sides.

Right-angled triangle

Contains a right angle.

Equilateral triangle

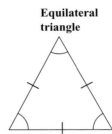

All angles equal, all sides equal.

Angle properties of quadrilaterals

A quadrilateral has four sides and its angles add up to 360°.

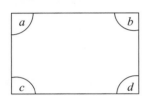

$$a + b + c + d = 360°$$

Here are some special quadrilaterals, together with their properties.

Square

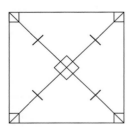

Properties
1 All angles equal 90°.
2 Opposite sides are parallel.
3 All sides are equal.
4 Diagonals are equal in length and bisect each other at right angles.

Rectangle

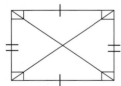

Properties
1 All angles equal 90°.
2 Opposite sides are parallel.
3 Opposite sides are equal.
4 Diagonals are equal in length and bisect each other.

Parallelogram

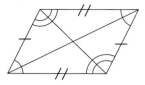

Properties
1 Opposite sides are parallel.
2 Opposite sides are equal.
3 Opposite angles are equal.
4 Diagonals bisect each other.

Trapezium

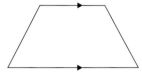

Properties
1 One pair of opposite parallel sides.

Rhombus

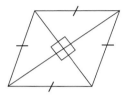

Properties
1 Opposite sides are equal.
2 Opposite sides are parallel.
3 Opposite angles are equal.
4 Diagonals cross at right angles.
5 Diagonals bisect each other.

Kite

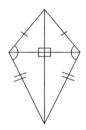

Properties
1 Adjacent sides are equal.
2 One pair of opposite angles is equal.
3 Diagonals cross at right angles.
4 One diagonal is bisected.

Angle properties of regular polygons

A polygon is a closed shape with straight sides. There are two types:

- A regular polygon has all sides equal and all angles equal.
- An irregular polygon has at least one side a different size from the others and at least one angle unequal.

Here is a regular pentagon, showing interior and exterior angles.

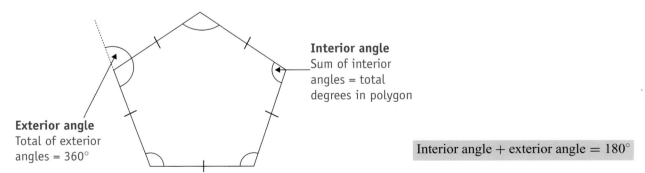

Interior angle
Sum of interior
angles = total
degrees in polygon

Exterior angle
Total of exterior
angles = 360°

Interior angle + exterior angle = 180°

For a regular polygon, the interior angles will be equal and the exterior angles will be equal. The sum of the interior angle and the exterior angle is $180°$.

To find the total degrees inside any regular polygon, we can use the formula:

Total degrees in a polygon = (Number of sides − 2) × $180°$
$$t = (n - 2) \times 180°$$

The common polygons have special names and angle properties. We need to know their names but we can use the rules above to find their angle properties.

Example 3 Find the total degrees, the interior angle and the exterior angle of a regular 8-sided polygon.

Using the formula for total degrees, $t = (n - 2) \times 180°$
$$t = (8 - 2) \times 180°$$
$$t = 1080°$$

Interior angle = total degrees ÷ number of sides
$$= 1080° \div 8$$
$$= 135°$$

Exterior angle = $180°$ − interior angle
$$= 180° - 135°$$
$$= 45°$$

Alternatively, the exterior angle can be found by using the following formula:

$$\text{exterior angle} = \frac{360}{\text{number of sides}}$$

If we apply these steps to the most common polygons we'll obtain the table below.

Number of sides	Name	Total degrees	Interior angle
3	Triangle	180	60
4	Quadrilateral	360	90
5	Pentagon	540	108
6	Hexagon	720	120
7	Heptagon	900	128.57 . . .
8	Octagon	1080	135
9	Nonagon	1260	140
10	Decagon	1440	144
11	Hendecagon	1620	147.27 . . .
12	Dodecagon	1800	150

Remember that a regular triangle is an equilateral triangle and a regular quadrilateral is a square.

Angle in a semicircle

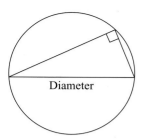

Any triangle using the diameter of a circle as its longest side is a right-angled triangle.

Angle between tangent and radius of a circle

A tangent is a straight line that just touches the circumference of a circle.

The circle below has two tangents at its circumference.

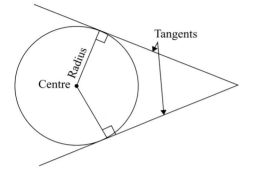

The angle formed between the tangent and the radius of a circle is 90°.

29 | Angle properties

Angle properties of irregular polygons, angles in a circle (circle theorems)

Angle properties of irregular polygons

Although irregular polygons do not have all equal angles or all equal sides, the interior and exterior angles still sum to $180°$.

Example 1 | For the polygon below find the sum of the interior angles and the size of angle x.

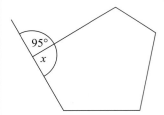

The polygon in the diagram has 5 sides, so it is an irregular pentagon. Using the formula to find the angle sum:

$$t = (n - 2) \times 180°$$
$$t = (5 - 2) \times 180°$$
$$t = 540°$$

To find x, we know that interior angle + exterior angle = $180°$.

So $x = 180° - 95°$
$\qquad x = 85°$

■ Note that we are not able to calculate the other four angles without more information, but we know that the sum of the remaining four angles is $540° - 85° = 455°$.

The permutations for this type of question are endless, so just remember the three basic formulae:

- $t = (n - 2) \times 180°$
- interior angle + exterior angle = $180°$
- sum of exterior angles = $360°$

Angle at the centre of a circle is twice the angle at the circumference

If the angle on the circumference is x degrees then the angle at the centre will be $2x$ degrees.

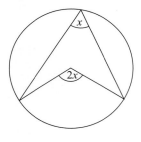
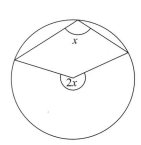
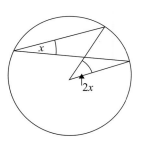

Example 2 | Find the values of x and y in the following diagram.

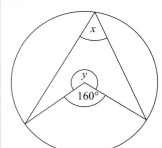

The angle at the centre is twice the size of the angle at the circumference.

So $\quad x = 160° \div 2 = 80°$
$\qquad y = 360° - 160° = 200°$ (angles at a point)

Angles in the same segment are equal

A segment is a part of a circle.

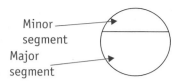

Minor segment

Major segment

The large part is called the **major segment** and the smaller part the **minor segment.**

Angles in the same segment are equal.

These angles are equal (same segment).

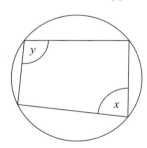

■ Note that there are two triangles formed and that they are similar to each other – all the angles in one triangle are equal to the angles in the other triangle.

Angles in opposite segments are supplementary

Remember that supplementary angles add up to $180°$.

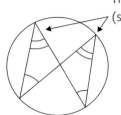

Angles x and y are opposite to each other and are supplementary:

$$x + y = 180°$$

The same will be true for the other two opposite angles.

■ Note that we can also say that the opposite angles of a cyclic quadrilateral are supplementary.

30 | Locus

Construct, using compasses and rulers, loci which are (a) a given distance from a fixed point, (b) a given distance from a straight line, (c) equidistant from two fixed points and (d) equidistant from two intersecting straight lines

A **locus** is the path traced by a moving object. The plural of locus is **loci**.

There are four types of locus we need to examine.

Loci which are at a given distance from a fixed point

An object moving at a fixed distance from a fixed point will always follow a **circular** path.

Example 1   Draw the locus of a point P which is always 3 cm from a fixed point X.

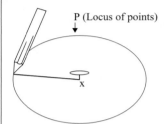

First mark point X on the paper, allowing at least 3 cm space around it.

Make the string of length 3 cm. Hold one end in place at point X, place a pencil at the other end and draw all the points around X which are 3 cm away.

Label the locus of points P.

Alternatively, we could have used compasses set to 3 cm and drawn a circle around X.

Loci which are at a given distance from a straight line

The locus of points at a fixed distance from a line is two parallel lines, one on either side of the line.

Example 2   Draw the locus of points P which are 2 cm from a straight line XY.

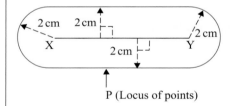

First we draw two parallel lines 2 cm either side of the fixed straight line XY.

Then, using compasses set to 2 cm at each end of the line, draw a 2 cm semicircle.

Join the parallel lines to the compass-drawn loci.

Loci which are equidistant from two fixed points

The easiest way to draw the locus of points which are an equal distance from two fixed points is to use a set of compasses. Draw a line connecting the two fixed points, then construct the perpendicular bisector of the line.

Example 3   Draw the locus of points P which are equidistant from two fixed points X and Y.

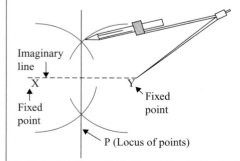

First join points X and Y.

Construct the perpendicular bisector of this line.

This perpendicular bisector is now the locus of points which are equidistant from points X and Y.

Loci of points which are equidistant from two given interesecting straight lines

The locus of points equidistant from two intersecting straight lines is the bisector of the angles between the lines.

Example 4   Draw the locus of points P which are equidistant from two intersecting straight lines AB and XY.

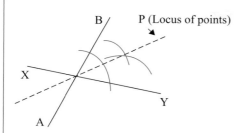

First bisect the angle between the two lines AB and XY (see Unit 27).

The angle bisector is the locus of points which are equidistant from the two intersecting lines.

Summary

- For loci of points from a fixed point X, use compasses at the required setting and draw a circle around X.
- For loci of points from a straight line XY, draw parallel lines either side of XY and use compasses for arcs.
- For loci of points from two fixed points X and Y, draw the perpendicular bisector of X and Y.
- For loci of points from two intersecting straight lines AB and XY, draw the bisector of the angles between AB and XY.

31 | **Mensuration**

Find the perimeter and area of a rectangle and a triangle, the circumference and area of a circle, the area of a parallelogram and a trapezium, the volume of a cuboid, prism and cylinder and the surface area of a cuboid and a cylinder

Perimeter and area of a rectangle and a triangle

The formulae for the area and perimeter are shown below.

Rectangle

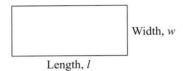

Width, w

Length, l

$$\text{Area} = \text{Length} \times \text{Width}$$
$$= l \times w$$
$$\text{Perimeter} = 2l + 2w$$

Triangle

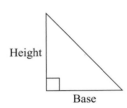

Height

Base

$$\text{Area} = \frac{\text{Base} \times \text{Height}}{2}$$
$$= \frac{b \times h}{2}$$
$$\text{Perimeter} = \text{sum of the lengths of three sides}$$

Example 1   Find the area and perimeter of the following shapes.

(a)

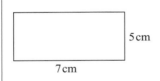

5 cm

7 cm

(b)

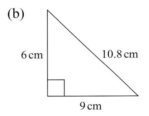

6 cm   10.8 cm

9 cm

(a)   Area = length × width
         = 7 cm × 5 cm
         = 35 cm²
Perimeter = $2l + 2w$
         = 2(7) + 2(5)
         = 24 cm

(b) $\text{Area} = \dfrac{\text{base} \times \text{height}}{2}$
         $= \dfrac{9\,\text{cm} \times 6\,\text{cm}}{2} = 27\,\text{cm}^2$
Perimeter = sum of three sides
         = 6 cm + 9 cm + 10.8 cm
         = 25.8 cm

Circumference and area of a circle

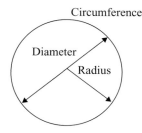

Circumference = distance around the circle
Diameter = distance across the centre
Radius = distance from centre to any
point on the circumference

Area of circle $= \pi r^2$
where $\pi = 3.142\ldots$, r = radius

Circumference of circle $= 2\pi r$ or πd
where d = diameter

| Example 2 | Find the area and circumference of a circle with a diameter of 8 cm. |

$$\text{Area} = \pi r^2 = 3.142 \times 4^2$$
$$= 3.142 \times 16$$
$$= 50.28 \text{ cm}^2 \qquad \text{Substitute the values given into the formulae. Note that we}$$
$$\text{Circumference} = 2\pi r \qquad\qquad \text{halve the diameter to find the radius.}$$
$$= 2 \times 3.142 \times 4$$
$$= 25.14 \text{ cm}$$

■ Note that to find the area of a semicircle with diameter of 8 cm, halve the answer above.

Area of a parallelogram and a trapezium

The formulae for the areas of these shapes are shown below.

Parallelogram

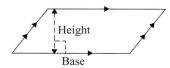

Area = Base × Height

Trapezium

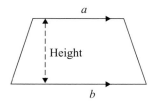

$$\text{Area} = \frac{a+b}{2} \times \text{height}$$
$$= \left(\frac{\text{sum of parallel sides}}{2} \right) \times \text{height}$$

| Example 3 | Find the area of the two shapes below. |

(a)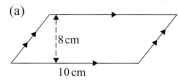

(b)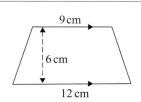

(a) Area = base × height
$= 10 \text{ cm} \times 8 \text{ cm}$
$= 80 \text{ cm}^2$

(b) Area $= \dfrac{a+b}{2} \times \text{height}$
$= \dfrac{9+12}{2} \times 6$
$= 10.5 \times 6$
$= 63 \text{ cm}^2$

Volume and surface area of a cuboid, triangular prism and cylinder

A prism is a solid with a constant cross-section. These are the important prisms.

Cuboid

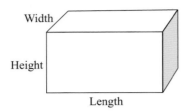

A **cuboid** is a prism with a rectangular base. An example would be a block of wood.

> Volume = Length × Width × Height
> Surface area = Total area of all six faces

Triangular prism

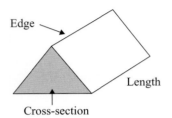

This solid has a triangular cross-section, so it is a **triangular prism**.

> Volume = Area of cross-section × Length
> Surface area = 2(area of cross-section) + area of other faces

Cylinder

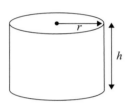

A cylinder is a prism with a circular cross-section. An example would be a can of coke.

> Volume = $\pi r^2 h$
> Surface area = $2\pi r(h + r)$

Example 4   The diagram shows the cross-section of a metal bar. If it is 12 cm long, find its volume.

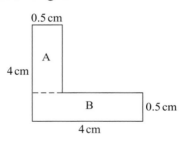

First, split the shape into two rectangles (see dotted line), then calculate the areas of each and add them.

$$\text{Area A} = 3.5\,\text{cm} \times 0.5\,\text{cm} = 1.75\,\text{cm}^2$$
$$\text{Area B} = 4\,\text{cm} \times 0.5\,\text{cm} \quad = 2\,\text{cm}^2$$
$$\text{Total} = 3.75\,\text{cm}^2$$

$$\text{Volume} = \text{Area of cross-section} \times \text{length}$$
$$= 3.75\,\text{cm}^2 \times 12\,\text{cm} = 45\,\text{cm}^3$$

Example 5   Find the total surface area of this cuboid.

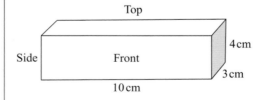

The cuboid has 6 faces, so we need to find the area of each and add them.

The front face and the hidden opposite face are each 10 cm by 4 cm rectangles.

The top and bottom faces are each 10 cm by 3 cm rectangles.

The side faces are each 4 cm by 3 cm rectangles.

$$\text{Area of front face} = 10\,\text{cm} \times 4\,\text{cm} = 40\,\text{cm}^2$$
$$\text{Area of top face} = 10\,\text{cm} \times 3\,\text{cm} = 30\,\text{cm}^2$$
$$\text{Area of side face} = 4\,\text{cm} \times 3\,\text{cm} \quad = 12\,\text{cm}^2$$
$$\text{So total surface area} = 40 \times 2 + 30 \times 2 + 12 \times 2 = 80 + 60 + 24 = 164\,\text{cm}^2$$

Example 6   The diagram below shows a laboratory flask which may be considered to be a sphere with a cylindrical neck. Calculate the volume of the flask, given that the volume of a sphere is $\frac{4}{3}\pi r^3$.

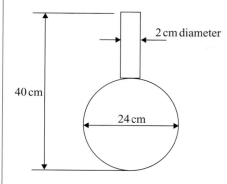

We have two basic shapes: a sphere of 24 cm diameter and a cylinder of 2 cm diameter. We need to calculate the volume of each and then add them.

$$\text{Cylinder length} = 40 \text{ cm} - 24 \text{ cm} = 16 \text{ cm}.$$

$$\text{Sphere volume} = \tfrac{4}{3}\pi r^3 = \tfrac{4}{3} \times 3.142 \times 12^3$$
$$= 7239.2 \text{ cm}^2$$

$$\text{Cylinder volume} = \pi r^2 h = 3.142 \times 1^2 \times 16$$
$$= 50.272 \text{ cm}^3$$

$$\text{Total volume} = 7239.2 + 50.272$$
$$= 7289.472 \text{ cm}^3$$

31 | Mensuration

Solve problems involving the arc length and sector area of a circle, the surface area and volume of a sphere, pyramid and cone (given formulae for the sphere, pyramid and cone)

Finding the arc length and sector area of a circle

The circle below illustrates an arc length and sector area.

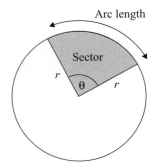

As the arc length is a fraction of the circumference of a circle the formula is:

$$\text{Arc length} = \frac{\theta}{360} \times 2\pi r$$

The sector area is a fraction of the total area of the circle, so the formula is:

$$\text{Sector area} = \frac{\theta}{360} \times \pi r^2$$

Here is an example on finding the arc length and sector area.

Example 1   Find the arc length and sector area of a circle with radius 8 cm and $\theta = 40°$.

$$\text{Arc length} = \frac{\theta}{360} \times 2\pi r = \frac{40}{360} \times 2 \times 3.142 \times 8$$
$$= 5.59 \, \text{cm}$$

$$\text{Sector area} = \frac{\theta}{360} \times \pi r^2 = \frac{40}{360} \times 3.142 \times 8^2$$
$$= 22.34 \, \text{cm}^2$$

■ If you are given the diameter instead of the radius, you need to halve the diameter. In other questions you will be given the arc length or sector area and asked to find the radius or the angle θ. In such cases you have to transpose the formula (see Unit 20).

Spheres, pyramids and cones

Cone

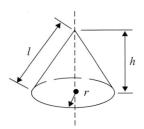

$$\text{Volume} = \tfrac{1}{3}\pi r^2 h$$

Frustum of a cone

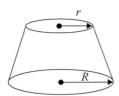

$$\text{Volume} = \tfrac{1}{3}\pi h(R^2 + Rr + r^2)$$

Pyramid

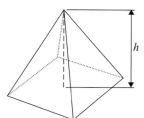

Volume $= \frac{1}{3}Ah$ where $A =$ area of base

Sphere

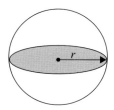

Volume $= \frac{4}{3}\pi r^3$

Surface area and volume of a sphere, pyramid and cone

The formulae will be given for these in the exam, but it is useful to look at some problems on this topic.

Example 2 A sphere has a diameter of 8 cm. Calculate its volume and surface area, given the formulae:

volume $= \frac{4}{3}\pi r^3$ surface area $= 4\pi r^2$ where r is the radius.

Volume $= \frac{4}{3} \times 3.142 \times 4^3$ Substitute the given values into the formulae.
$= 268.1 \text{ cm}^3$

Surface area $= 4 \times 3.142 \times 4^2$
$= 201.09 \text{ cm}^2$

Example 3 A pyramid has a square base of side 4 cm and a volume of 16 cm³. Calculate its height. (Volume of pyramid $= \frac{1}{3}Ah$, where A is the area of the base and h is the height.)

We have the volume and the area, so we need to rearrange the formula to find h.

$$V = \frac{1}{3}Ah, \text{ so } h = \frac{3V}{A}.$$

So $h = \dfrac{3 \times 16}{16} = 3$ cm. $A =$ area of the base $= 4\,\text{cm} \times 4\,\text{cm} = 16\,\text{cm}^2$

Example 4 A cone has a diameter of 7 cm and a volume of 308 cm³. Taking $\pi = 3.142$, calculate the vertical height of the cone. (Volume of cone $= \frac{1}{3}\pi r^2 h$, where r is the radius of the base and h is the height.)

We have the diameter and the volume, so we need to rearrange the formula to find h.

$$V = \frac{1}{3}\pi r^2 h, \text{ so } h = \frac{3V}{\pi r^2}.$$

So $h = \dfrac{3 \times 308}{3.142 \times 3.5^2} = 24$ cm.

32 | Trigonometry

Use bearings; use trigonometric ratios to find missing sides, angles; use Pythagoras' theorem

Revision on bearings

Bearings are angular directions. They are measured from due north in a clockwise direction and are written as three figures.

Example 1 Point B lies on a bearing of 095° from point A. Construct the bearing diagram.

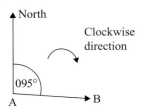

The starting point is A, since the question gives the bearing of B from A. The angle of 95° is drawn using a protractor, from due north, in a clockwise direction.

Note how we have written the angle as three figures and shown the direction arrows.

■ When we have bearings of 100° or more, these are already in three figures, so there is no need to add a zero.

Example 2 A ship sails from port P on a bearing of 120° for 10 km to a point X and then changes direction to bearing 070° to a point Y which is 8 km away. Construct an accurate scale diagram to show the ship's journey and hence find the shortest distance from the port to point Y.

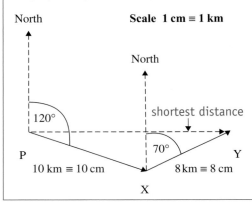

Scale 1 cm ≡ 1 km

Starting at P, we construct a north, then measure 120° with a protractor and draw a line of length 10 cm in this direction towards X.

At point X, we construct another north and measure 70° and draw a line of length 8 cm towards Y.

This is a representation of the journey drawn to scale.

Measure the distance PY with a ruler. It is 16 cm.

Distance PY = 16 km.

Back-bearings

Although not specifically mentioned in the syllabus statement, it is important to know what these are and how to calculate them. A back-bearing is the reverse of a bearing, so if A is on a bearing from B, then the bearing of B from A is the back-bearing.

Use the rule below to calculate a back-bearing from a given bearing.

> If the original bearing is less than 180°, add 180° to get the back-bearing.
> If the original bearing is greater than 180°, subtract 180° to get the back-bearing.

If A is on a bearing of 055° from B, what is the back-bearing of B from A?

As 055° is less than 180° we simply add 180° to it:

$$055° + 180° = 235°$$

We can show this diagrammatically.

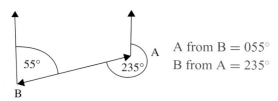

A from B = 055°
B from A = 235°

Back-bearings are used extensively in geography to calculate the compass headings when returning from a journey, and in the shipping and aeronautical industries.

Trigonometric ratios

We can use the trigonometric ratios to find the missing side(s) or angle(s) of any right-angled triangle, provided we have sufficient information.

The sides of a right-angled triangle are shown below. The **hypotenuse** is the longest side (always opposite the right angle), the **opposite** is the side opposite the angle x and the **adjacent** is the side next to the angle x. Remember their names and the corresponding ratios.

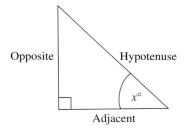

$$\text{Sine } x° = \frac{\text{Opposite}}{\text{Hypotenuse}}$$

$$\text{Cos } x° = \frac{\text{Adjacent}}{\text{Hypotenuse}}$$

$$\text{Tan } x° = \frac{\text{Opposite}}{\text{Adjacent}}$$

We often use the word SOHCAHTOA to remember the ratios:

s	o	h	c	a	h	t	o	a
i	p	y	o	d	y	a	p	d
n	p	p	s	j	p	n	p	j

Using the trigonometric ratios to find missing angles and sides

First familiarise yourselves with the following keys on your scientific calculator:

| Sin | Sin⁻¹ | Cos | Cos⁻¹ | Tan | Tan⁻¹ |

These are the trigonometric functions. The inverse functions, $\sin^{-1}$, $\cos^{-1}$ and $\tan^{-1}$, are usually obtained by pressing the [inv] or second function key on your calculator.

Finding a missing angle

In order to calculate a missing angle, we need to know the length of at least two sides. We can then name the sides accordingly and choose the appropriate ratio.

Find angle x in the triangle below.

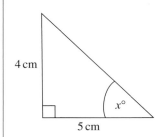

We can see from the triangle that the two given sides are the opposite and the adjacent, so we have to use the tan ratio.

$$\tan x = \frac{\text{opposite}}{\text{adjacent}} = \frac{4}{5} = 0.8$$

$$\tan^{-1} 0.8 = 38.7° \text{ (to 1 d.p.)}$$

To find the angle from the ratio, press the [Tan⁻¹] key.

| **Example 5** | Find angle y in the triangle below. |

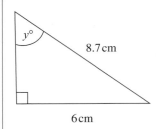

This time the two given sides are the hypotenuse and the opposite, so we have to use the sin ratio. (Note how the triangle has turned round but this does not affect the naming of the sides.)

$$\sin y = \frac{\text{opposite}}{\text{hypotenuse}} = \frac{6}{8.7} = 0.6897$$

$$\sin^{-1} 0.6897 = 43.6° \text{ (to 1 d.p.)}$$

■ If we were asked to find the remaining angle in the triangle, we would subtract the right angle and the calculated angle from $180°$. So in this case the remaining angle is $180 - 90 - 43.6 = 46.4°$.

Finding a missing side

In order to calculate a missing side we need the angle x and one other side.

(If we had two of the sides then there would be no need to use trigonometry, we could use Pythagoras' theorem to calculate the remaining side.)

| **Example 6** | Find the length of sides y and z in the triangle below. |

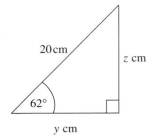

As we have the hypotenuse and the angle, we could use either the sine or cosine ratio.

Using the sine ratio:

$$\sin 62° = \frac{\text{opposite}}{\text{hypotenuse}} = \frac{z}{20}$$

Rearranging for z:

$$z = \sin 62° \times 20 \, \text{cm} = 17.7 \, \text{cm}$$

To find y we can use Pythagoras' theorem as we now have two of the sides:

$$z = \sqrt{20^2 - 17.7^2} = 9.3 \, \text{cm}$$

| **Example 7** | Find x in the triangle below. |

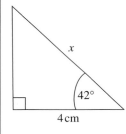

In this question, the unknown side is the hypotenuse while the known side (4 cm) is adjacent to the known angle. Therefore we have to use the cosine ratio:

$$\cos 42° = \frac{\text{adjacent}}{\text{hypotenuse}} = \frac{4 \, \text{cm}}{x}$$

Rearranging for x:

$$x = \frac{4 \, \text{cm}}{\cos 42°} = 5.4 \, \text{cm}$$

Example 8

Two boats X and Y, sailing in a race, are shown in the diagram below. Boat X is 145 metres due north of buoy B. Boat Y is due east of buoy B. Boats X and Y are 320 metres apart. Calculate:

(a) the distance BY
(b) the bearing of Y from X
(c) the bearing of X from Y.

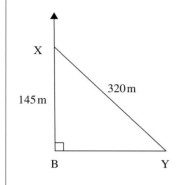

(a) Distance BY can be found using Pythagoras' theorem:

$$BY = \sqrt{320^2 - 145^2} = 285.3 \text{ metres}$$

(b) First find $\widehat{X}$ and then subtract it from 180 to find the bearing of Y from X

$$\cos \widehat{X} = \frac{\text{adjacent}}{\text{hypotenuse}} \qquad = \frac{145}{320} = 0.4531$$

$\cos^{-1} 0.4531 = 63°$, so the bearing $= 180° - 63° = 117°$

(c) The bearing of X from Y is the back-bearing of Y from $X = 117° + 180° = 297°$.

32 | Trigonometry

Use trigonometry to solve problems in two and three dimensions involving angles of elevation and depression; extend sine and cosine functions to angles between 90° and 360°; solve problems using the sine and cosine rules for any triangle and use area of triangle $= \frac{1}{2}ab \sin C$

Angles of elevation and depression

Angle of elevation

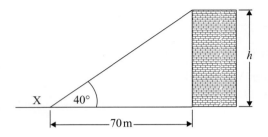

The angle of elevation is the angle above the horizontal through which a line of sight is raised.

The diagram shows a tower 70 metres away from a point X on the ground. If the angle of elevation of the top of the tower from X is 40°, calculate the height of the tower.

$$\tan 40° = \frac{h}{70}$$
$$\text{so } h = \tan 40° \times 70$$
$$= 58.7 \text{ metres} \ (\text{to } 1 \text{ d.p.})$$

Angle of depression

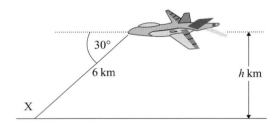

The angle of depression is the angle below the horizontal through which a line of sight is lowered.

The diagram shows an aeroplane receiving a signal from point X on the ground. If the angle of depression of point X from the aeroplane is 30°, calculate the height at which the plane is flying.

$$\sin 30° = \frac{h}{6}$$
$$\text{so } h = \sin 30° \times 6$$
$$= 3.0 \text{ km}$$

Angles between 0° and 360°

When calculating the size of angles using trigonometry, there are often two solutions between 0° and 360°. Most calculators, however, will only give the first solution. In order to calculate the value of the second possible solution, an understanding of the shape of the sine and cosine curves is required.

The sine curve

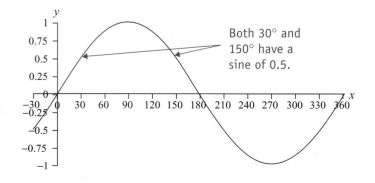

Both 30° and 150° have a sine of 0.5.

This is the graph of $y = \sin x$, where x is the size of the angle in degrees. We can see from the graph that:

- it has a period of 360° (repeats itself every 360°)
- it has a maximum value of 1
- it has a minimum value of -1.

If we take, for example, sin 30° = 0.5, we notice that the sine of 150° is also 0.5, although this second value would not be shown on our calculators.

The cosine curve

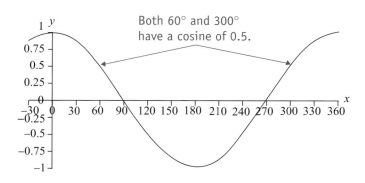

Both 60° and 300° have a cosine of 0.5.

This is the graph of $y = \cos x$, where x is the size of the angle in degrees. We can see from the graph that:

- it has a period of 360°
- it has a maximum value of 1
- it has a minimum value of −1.

If we take, for example, cos 60° = 0.5, we notice that the cosine of 300° is also 0.5. Again this second value would not be shown on our calculators.

The sine and cosine rules

These are extremely powerful rules as they can be used for any type of triangle, not just a right-angled triangle. The basic notation used for the triangle is shown below.

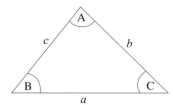

Angle A is opposite to side a.
Angle B is opposite to side b.
Angle C is opposite to side c.

The sine rule

The sine rule is used when we are given:
- one side and any two angles, *or*
- two sides and an angle opposite to one of the sides.

$$\frac{a}{\sin A} = \frac{b}{\sin B} = \frac{c}{\sin C}$$ This can be transposed so that the angles are on the top of the fractions.

Example 1   Calculate the length of side BC.

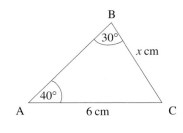

The sine rule has to be used because it is not a right-angled triangle.

Side BC = a.
Using the formula $\dfrac{a}{\sin A} = \dfrac{b}{\sin B}$

$$a = \frac{\sin 40° \times 6\,cm}{\sin 30°}$$
$$= 7.7\,cm \ \text{(to 1 d.p.)}$$

■ We did not need to use sin C, because angle B and side b were available.

Example 2   Calculate the size of angle C.

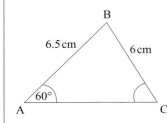

We are looking for an angle, so we need to invert the formula:

$$\frac{\sin C}{c} = \frac{\sin A}{a}$$ Note that we do not have sin B.

$$\sin C = \frac{6.5\,\text{cm} \times \sin 60°}{6\,\text{cm}}$$

$$\sin^{-1} 0.94 = 69.8° \text{ (to 1 d.p.)}$$

■ We did not have angle B or side b to use, but angle C and side c were sufficient.

The cosine rule

The cosine rule is used when we are given: • two sides of a triangle and the angle between them, *or*
• three sides of a triangle

$$a^2 = b^2 + c^2 - 2bc \cos A$$

or

$$\cos A = \frac{b^2 + c^2 - a^2}{2bc}$$

Example 3   Calculate the size of angle A.

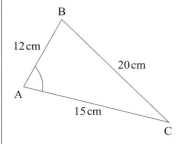

Using the formula to find angle A,

$$\cos A = \frac{15^2 + 12^2 - 20^2}{2 \times 15 \times 12}$$

$$= -0.086$$

$$\cos^{-1} -0.086 = 94.9° \text{ (to 1 d.p.)}$$

■ We could have found any angle using the formula, we just have to rearrange it for the required angle.

Example 4   Calculate the length of side AB.

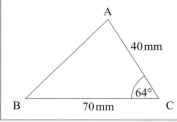

Side AB is side c of the triangle, so we have to rearrange the formula for c:

$$c = \sqrt{a^2 + b^2 - 2ab \cos C}$$

$$= \sqrt{70^2 + 40^2 - 2 \times 70 \times 40 \times \cos 64°}$$

$$= 63.60\,\text{mm}$$

You only need the minimum amount of information to be able to calculate all of the sides and all of the angles of a triangle. It is always worth doing a check. Remember that the longest side is always directly opposite the largest angle, and the shortest side is opposite the smallest angle.

Area of a triangle

Area of $\triangle = \dfrac{1}{2}ab\sin C$	where a and b are two sides and angle C is the angle between them. This formula can be rearranged for any angle.

Example 5   Find the area of the triangle below.

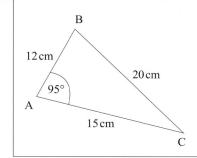

Angle A is known, so the formula becomes:

$$\text{Area} = \frac{1}{2}bc\sin A$$
$$= \frac{1}{2} \times 15 \times 12 \times \sin 95°$$
$$= 89.7\,\text{cm}^2$$

In some questions we are given the area of the triangle and some other information and asked to find either a missing side or the missing angle between the sides. This is straightforward provided we can rearrange formulae (see Unit 20).

Trigonometry in three dimensions

Three-dimensional shapes are **solid shapes**. In order to calculate angles and sides in solid shapes we need to understand the terms **plane** and **angle between a line and a plane**.

The plane

A plane is a surface such as the top of a table or the cover of a book. A plane has two dimensions.

The angle between a line and a plane

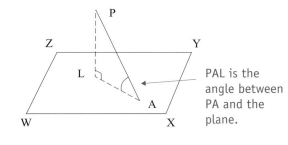

PAL is the angle between PA and the plane.

The line PA intersects the plane WXYZ at A.

To find the angle between PA and the plane, draw PL perpendicular to the plane and join AL.

The angle PAL is now the angle between the line and the plane.

Example 6   The diagram below shows a square-based pyramid of side 8 cm. The height of the pyramid is 12 cm. Calculate   (a) EF   (b) angle VEF   (c) VE (slant height)   (d) area of $\triangle$VAD

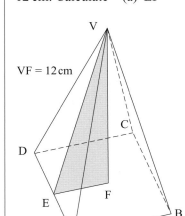

(a) EF = base of triangle VEF   This is half way across
       = 4 cm   the square of side 8 cm.

(b) Angle VEF $= \tan^{-1}\dfrac{12}{4}$ Angle formed between
       $= 71.6°$ line and plane.

(c) VE $= \sqrt{12^2 + 4^2}$ Using Pythagoras.
       $= 12.6\,\text{cm}$

(d) Area $\triangle$VAD $= \dfrac{1}{2} \times \text{base} \times \text{height}$
       $= \dfrac{1}{2} \times 8 \times 12.6$
       $= 50.4\,\text{cm}^2$

Example 7   The diagram below shows a pyramid on a rectangular base. Calculate the length VA and the angle VAE.

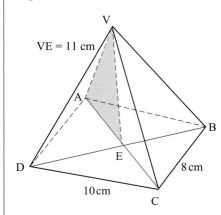

To calculate VA we can use Pythagoras:

$$VA = \sqrt{VE^2 + AE^2}$$

First we need to calculate AE. AE is half the distance of AC, so we can say:

$$AE = \frac{1}{2}\sqrt{10^2 + 8^2}$$

$$= 6.4\text{ cm}$$

$$\text{So } VA = \sqrt{11^2 + 6.4^2} = 12.7\text{ cm}$$

$$\text{Angle } VAE = \tan^{-1}\frac{11}{6.4} = 59.8°$$

33 | Statistics

Interpret, tabulate and draw bar charts, pictograms, pie charts, histograms (equal intervals) from data; calculate mean, median and mode and distinguish between purposes for which they are used; *scatter diagrams (including line of best fit by eye), positive, negative and zero correlation*

The italicised areas of study are not examined until the first examination in 2006.

The word '**statistics**' means 'gathering and displaying of information (data)'. We have all seen the use of statistics in our world, e.g. population growths, temperature/rainfall charts, and we should be able to draw simple conclusions from them. When we conduct our own statistical surveys we need to follow some simple procedures so that the information is first collected, then ordered and finally displayed in a suitable fashion.

Collect, classify and tabulate statistical data

This refers to the actual collection and ordering of the data we require. To collect the data we could write a questionnaire and ask people to answer the relevant questions, or we could ask people directly and complete the results ourselves, provided that the sample size is small.

Example 1

In a recent mathematics test involving 50 students, the following results were obtained:

40	30	50	50	60	50	80	70	60	70	80	30
90	50	40	10	80	70	50	60	60	70	50	20
50	20	60	90	50	70	60	50	60	20	80	60
70	30	30	80	70	60	50	50	60	40	30	40
50	70										

Classify and tabulate the data.

We can order the data by using a **frequency distribution**.

Score	Tally	Frequency
10	I	1
20	III	3
30	IIII	5
40	IIII	4
50	IIII IIII II	12
60	IIII IIII	10
70	IIII III	8
80	IIII	5
90	II	2
—	—	Total = 50

■ Note how the data was placed in numerical order and the total then checked to ensure that we had accounted for all 50 students.

We can also see from the table that most students' scores are in the 50 to 70 range, which is what you'd expect from a well-constructed test.

Bar charts, pie charts and pictograms

It is easier to draw conclusions if data is displayed in chart form.

The example that follows shows the different ways we can do this.

Example 2 Ninety people were asked for their favourite flavour of crisps. The results were as follows:

Plain	Cheese and onion	Salt and vinegar	Chicken	Tomato
20	25	18	15	12

Show the information as   (a) a bar chart   (b) a pie chart   (c) a pictogram.

(a)

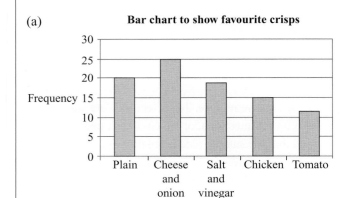

With a bar chart, the number of crisps or **frequency** goes on the vertical axis while the **category** or data goes on the horizontal axis.

(b)   **Pie chart to show favourite crisps**

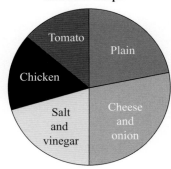

This time the data is shown as sectors of a circle.

The larger the circle sector, the greater the number of students who prefer that flavour.

We can use the following formula to calculate the sector angle:

$$\text{angle} = \frac{\text{amount} \times 360}{\text{total}}$$

$$\text{Plain} = \frac{20 \times 360}{90} = 80°$$

$$\text{Cheese and onion} = \frac{25 \times 360}{90} = 100°$$

$$\text{Salt and vinegar} = \frac{18 \times 360}{90} = 72°$$

$$\text{Chicken} = \frac{15 \times 360}{90} = 60°$$

$$\text{Tomato} = \frac{12 \times 360}{90} = 48°$$

■ When you have the sector angles, use your protractor to draw the angles, using the top as the starting point.

(c) A pictogram to show favourite crisps

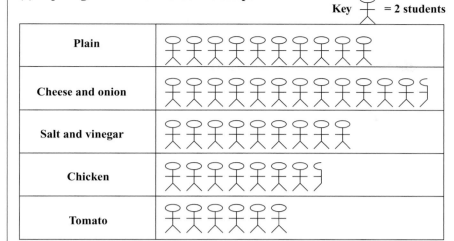

Plain		$10 \times 2 = 20$
Cheese and onion		$12.5 \times 2 = 25$
Salt and vinegar		$9 \times 2 = 18$
Chicken		$7.5 \times 2 = 15$
Tomato		$6 \times 2 = 12$

With a pictogram, the amount in each category is shown as a symbol or picture. In this case we are showing students, so we can use matchstick men to represent every 2 students.

■ For the Cheese and onion and Chicken flavours we have shown half a picture to represent one person. Every pictogram must have a suitable key, so we can read the data correctly.

Histograms with equal intervals

A **histogram** shows the frequency of the data in the form of bars; the area of the bars gives the number of items in the class interval. If all the class intervals are the same, then the bars will be the same width and frequencies can be represented by the heights of the bars.

So if we now show the data from Example 1 as a histogram it would look like this:

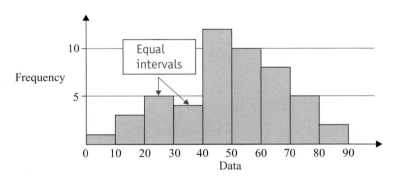

As the intervals are the same size, i.e. every 10 marks, then the change in height will represent the actual change in area.

■ In the core syllabus a histogram is virtually identical to a bar chart except that the bars are always joined.

Grouped class intervals

Each class interval includes a range of values.

| Example 3 | Draw a histogram from the table which shows the distribution of heights of 21 students in a mathematics class. The measurements were made to the nearest centimetre. |

Class interval	120–129	130–139	140–149	150–159	160–169
Frequency	2	4	7	5	3

The class boundaries are:

119.5   129.5   139.5   149.5   159.5   169.5

Area of rectangle = class width × height of rectangle   Class width = 10 cm for each class, e.g. 120–129.

Frequencies will be on the vertical axis and class intervals on the horizontal axis.

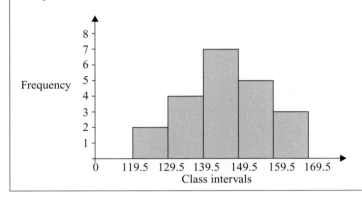

Area is proportional to the height of the bars.

All class interval widths are the same.

Use graph paper to ensure accuracy.

Mean, median and mode

These are all types of averages. The word 'average' usually means somewhere in the middle, but there are three types of statistical averages in common use today. Depending on the circumstances where they are used, each one has an advantage over the others.

Mean

This is the most commonly used type of average. It is used to give an accurate arithmetic middle value of a group of figures, e.g. finding the average height of a population.

$$\text{Mean} = \frac{\text{Total of all the values}}{\text{Number of values}}$$

Median

This is a lesser used type of average. It is useful when you want to find the middle value of a set of numbers which contain extreme values.

Median = Middle value when numbers are in ascending order

Mode

This is useful when you need to know which value occurs most often.

> Mode = Value which occurs most frequently

This example shows how we calculate each type of average and their relative advantages.

Example 4 A survey was done on the shoe sizes of 25 pupils. The results were as follows:
39, 40, 40, 44, 44, 40, 40, 41, 41, 44, 42, 42, 45, 45, 41, 41, 41, 42, 42, 43, 38, 39, 39, 43, 45.
Find the mean, median and mode of the results.

$$\text{Mean} = \frac{\text{total of values}}{\text{number of values}} = \frac{39 + 40 + 40 + 44 + 44 + 40 + 41 + \cdots + 43 + 45}{25} = 41.64$$

Median: 38, 39, 39, 39, 40, 40, 40, 40, 41, 41, 41, 41, 41, 42, 42, 42, 42, 43, 43, 44, 44, 44, 45, 45, 45.

| ↑ | ↑ | ↑ |
| 1st value | Median value = 41 (13th value) | 25th value |

First we place the numbers in ascending order. Then we count along to the value which is in the middle, in this case the 13th value.

Median = 41

Mode = most frequent value = 41   This value occurs most often.

■ The mean would be the most useful if you wanted to find an accurate middle value and then conclude who has bigger or smaller feet than average.

If a shoe shop wanted to keep adequate stocks of shoes of the most commonly requested size, it would want to know the median and the mode.

Example 5 The heights, in centimetres, of 10 students are: 152, 155, 153, 162, 148, 149, 152, 154, 155, 152.
Find the mean, median and mode.

$$\text{Mean} = \frac{\text{total of values}}{\text{number of values}} = \frac{152 + 155 + 153 + 162 + 148 + 149 + 152 + 154 + 155 + 152}{10}$$
$$= 153.2 \text{ cm}$$

Median: 148, 149, 152, 152, 152, 153, 154, 155, 155, 162

Numbers placed in ascending order. The middle number is between the 5th and 6th number, so we take the average of these two values.

Median = 152.5 cm   The middle value of 152 and 153.

Mode = most frequent value = 152 cm   This value occurs most often.

Scatter diagrams

Scatter diagrams (or scatter graphs) are used to see if there are any relationships between two sets of data. The data are plotted as points on a graph and if these points tend to lie in a straight line, we can say there is a relationship or **correlation** between them.

Types of correlation

There are three types:
- positive correlation
- negative correlation
- no correlation

	Positive correlation		Negative correlation		No correlation

Positive correlation

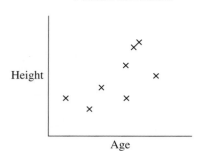

As age increases so does height.

Negative correlation

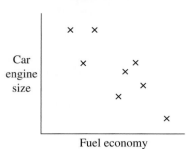

As engine size increases, the fuel economy decreases.

No correlation

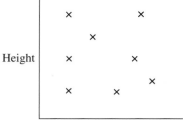

There is no link between Geography mark and height.

Lines of best fit

Example 6

The table below shows the percentage marks gained in a recent Physics test and an English test by the same 14 students.

Physics	47	57	64	50	70	80	67	62	76	90	83	93	90	76
English	43	46	51	55	55	65	65	67	69	72	74	78	79	79

We will now plot a graph of the data and construct a **line of best fit**, which should go through the middle of all the points:

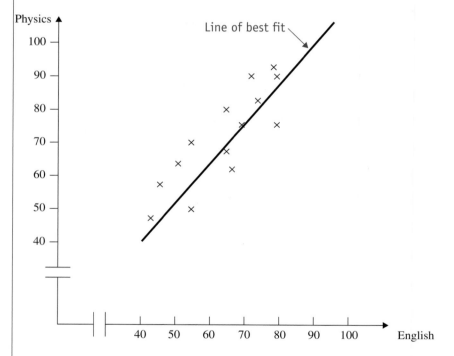

■ We can see from the line of best fit that there is a positive correlation between the Physics marks and the English marks.

We can therefore reasonably conclude that a student who scores well in Physics is likely to also score well in English.

Apart from using lines of best fit to give a clearer indication as to what type of correlation exists (if any), we can also use them for estimating a value not given in the table.

Example 7 Example 6 showed the percentage marks gained in recent Physics and English tests by 14 students. If a 15th student took the Physics test scoring 61% but missed the English test, write down a reasonable estimate of the score he or she would have obtained for English.

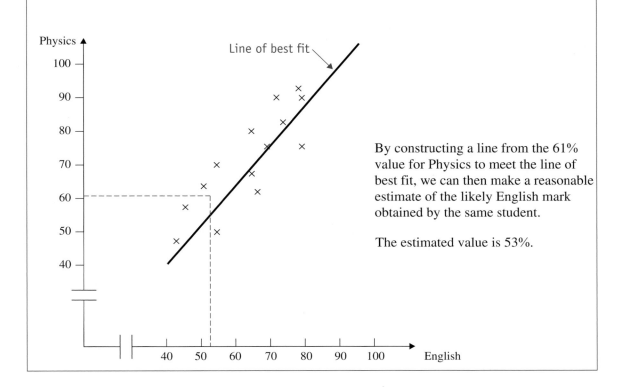

By constructing a line from the 61% value for Physics to meet the line of best fit, we can then make a reasonable estimate of the likely English mark obtained by the same student.

The estimated value is 53%.

33 | Statistics

Construct histograms (unequal intervals), cumulative frequency diagrams; estimate median, upper/lower quartiles, interquartile range; estimate the mean; find modal class and median from a grouped frequency distribution

Histograms with unequal intervals

We know that a histogram shows the frequency in terms of the bar areas. When we have equal intervals each bar width will be equal and hence the area is proportional to the height of the bar. However, when the intervals are unequal, we have to find the height of the bar in another way so that it is the area of the bar that shows the frequency.

A histogram with unequal intervals is shown below.

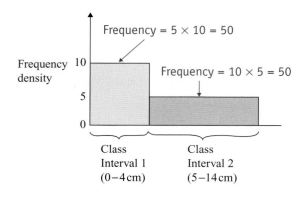

The height of the bar is now called the **frequency density** and it is found by the following formula:

$$\text{Frequency density} = \frac{\text{Frequency}}{\text{Class interval}}$$

From the diagram we can see that, although the frequency densities are different, the actual areas (frequencies) are the same. Although the height of the first bar is twice the height of the second, the width of the second bar is twice the width of the first.

Ideally, histograms should be drawn on graph paper with each axis to a specified scale so that we can compare the areas accurately. Example 1 shows a practical example of a histogram.

Example 1   The lengths of 21 seedlings were measured and the results recorded in the table below.

Draw a histogram to display this data and comment on which class interval had the greatest number of seedlings.

Height (cm)	Frequency
$0 \leq h < 1.0$	6
$1.0 \leq h < 1.5$	3
$1.5 \leq h < 2.0$	4
$2.0 \leq h < 2.25$	3
$2.25 \leq h < 2.5$	5

We know the frequency and the class interval but we need to calculate the height of the bars (frequency density), so we'll use the formula

$$\text{Frequency density} = \frac{\text{Frequency}}{\text{Class interval}}$$

By adding another column on the table to include frequency density, we can then plot the histogram.

Height (cm)	Frequency	Frequency density
$0 \leq h < 1.0$	6	$6 \div 1 = 6$
$1.0 \leq h < 1.5$	3	$3 \div 0.5 = 6$
$1.5 \leq h < 2.0$	4	$4 \div 0.5 = 8$
$2.0 \leq h < 2.25$	3	$3 \div 0.25 = 12$
$2.25 \leq h < 2.5$	5	$5 \div 0.25 = 20$

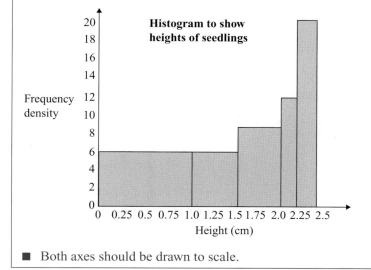

In the histogram the frequencies are now represented by the areas of the bars, e.g. there are 3 seedlings with a height of 1.0–1.5 cm because the area of the bar is $0.5 \times 6 = 3$.

The class interval with the greatest number of seedlings is 0–1.0 cm as this has the greatest area.

■ Both axes should be drawn to scale.

Cumulative frequency diagrams

A cumulative frequency diagram is an alternative to a histogram for presenting a frequency distribution. It is a useful way to find the median and quartiles (upper and lower quarters) of a distribution, and also for determining if the data is fair. A fair distribution will always take the typical shape below.

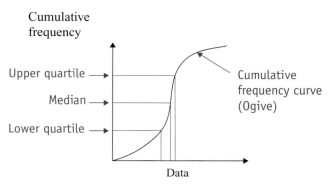

When plotting a cumulative frequency curve we must follow certain steps:

1. In the table add another column (or row) for cumulative frequency next to the frequency figures.
2. Plot the graph of cumulative frequency figures against the upper boundary of each class interval.
3. Use the cumulative frequency curve to find the median and percentiles.

Median is the 50% value of cumulative frequency (using graph, read off the value).

Upper quartile is the 75% value of cumulative frequency (as above, construct line from the graph). The upper quartile is the value of data corresponding to 75% of the total frequency.

Lower quartile is the 25% value of cumulative frequency (construct line again). The lower quartile is the value of the data corresponding to 25% of the total frequency.

Interquartile range = upper quartile − lower quartile (the middle 50% of the data).

Example 2 The table below shows the distribution of the scores obtained by 200 pupils in a maths test. Draw a cumulative frequency curve and hence find:

(a) median   (b) upper and lower quartiles   (c) interquartile range

Score	26–30	31–35	36–40	41–45	46–50	51–55	56–60	61–65	66–70	71–75	76–80
Pupils (frequency)	3	6	10	21	38	55	32	19	9	5	2
Cum. frequency	3 (0 + 3)	9 (3 + 6)	19 (9 + 10)	40 (19 + 21)	78 (40 + 38)	133	165	184	193	198	200

The first step is to include an extra row in the table for cumulative frequency, which is a running total of frequencies. Note how the final figure is 200, which corresponds to the 200 people in the question.

Plot the graph of cumulative frequency against the upper boundary of each class interval.

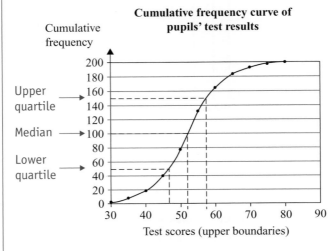

Cumulative frequency curve of pupils' test results

Cumulative frequency

Upper quartile →

Median →

Lower quartile →

Test scores (upper boundaries)

From the graph we find:

(a) Median (50% of 200 = 100)
    = 53 (approx)

(b) Upper quartile = 75% value
    = 58 (approx)

    Lower quartile = 25% value
    = 46 (approx)

(c) Interquartile range = 58 − 46
    = 12

■ The axes should be drawn to scale so that the answers obtained are as accurate as possible.

Calculating an estimate of the mean for grouped and continuous data

We have looked at calculating the mean for individual and discrete data. When we have grouped data the method is slightly different:

$$\text{Mean (estimate)} = \frac{(\text{Mid-value of class intervals} \times \text{Frequencies})}{\text{Total frequency}}$$

For each class interval we have to find the mid-value and multiply by its frequency. We then find the sum of these products and finally divide by the total frequency.

Example 3	In a chemistry experiment, the weights (grams) of 100 grains of salt were found. The distribution of weights is given in the following table. Calculate an estimate of the mean.

Mid-value of
0 and 4 = 2
↓

Mid-value of 20
and 24 = 22
↓

Weight (g)	0–4	5–9	10–14	15–19	20–24	25–29	30–34	35–39	40–44
Frequency	1	2	15	21	26	28	4	2	1

$$\text{Mean (estimate)} = \frac{(2 \times 1) + (7 \times 2) + (12 \times 15) + (17 \times 21) + (22 \times 26) + (27 \times 28) + (32 \times 4) + (37 \times 2) + (42 \times 1)}{100}$$

$$= \frac{2 + 14 + 180 + 357 + 572 + 756 + 128 + 74 + 42}{100}$$

$$= 21.25$$

■ This is only an estimate because we don't know exactly how the numbers were arranged in each class interval. For example, in the 10–14 gram interval we know there were 15 salt grains but they could have been anywhere in that range (near the lower boundary or higher boundary).

Finding the modal class of a grouped distribution

As mentioned in the core section, the mode is the number which occurs most frequently, so if we have a grouped distribution, the interval with the highest frequency is the modal class.

Look at the table of results from the previous example:

Weight	0–4	5–9	10–14	15–19	20–24	25–29	30–34	35–39	40–44
Frequency	1	2	15	21	26	28	4	2	1

↑
The modal class is 25–29 since this has the highest frequency.

Finding the median from a grouped frequency distribution

This question is usually asked in relation to individual data, so use the table of results below as an example.

A survey of number of children in 17 families

Children	0	1	2	3	4	5
Frequency	2	3	5	4	2	1

Running total   2   5   10   14   16   17

We know that the median is the middle value when the data is in ascending order, but in a frequency diagram the data is already in the correct order, so we just need to find the middle value.

A quick method of finding the middle value is to use the formula:

$$\text{Middle value} = \frac{\text{Total frequency} + 1}{2}$$

So median $= \dfrac{17 + 1}{2} = $ 9th value $= 2$ children

34 | Probability

Calculate the probability of a single event as either a fraction or a decimal (not a ratio), *understand and use the probability scale from 0 to 1; probability of events not occuring, relative frequency*

The italicised areas of study are not examined until the first examination in 2006.

Probability is the study of chance, or the likelihood of an event happening. We can find the probability of an event as a numerical value by using the formula:

$$\text{Probability} = \frac{\text{Number of favourable outcomes}}{\text{Total number of possible outcomes}}$$

 For example, consider a dice numbered 1, 2, 3, 4, 5, 6 as shown in the diagram. We want to find the probability of throwing a number 6 with one throw.

Using the formula,

$$\text{Probability} = \frac{\text{Number of favourable outcomes}}{\text{Total number of possible outcomes}}$$

$$P(6) = \frac{1}{6}$$

← 6 can only occur once

← There are 6 possible outcomes

The answer $\frac{1}{6}$ could also be shown as a decimal $1 \div 6 = 0.166\ 666\ldots$

Simplify your answers, if possible, and give the final answer as required, as either a fraction or a decimal.

Example 1 From one throw of a dice find the probability of getting:
(a) an even number
(b) a multiple of 3
(c) a number 7

(a) $P(\text{even}) = \frac{3}{6} = \frac{1}{2}$ There are 3 even numbers (2, 4 and 6), out of 6 possible outcomes.

(b) $P(\text{multiple of 3}) = \frac{2}{6} = \frac{1}{3}$ There are 2 multiples of 3 (3 and 6), out of 6 possible outcomes.

(c) $P(7) = 0$ Number 7 is not possible.

The probability scale

Probabilities can be expressed on a scale as shown below:

Impossible	Evens	Certain
0	$0.5/\frac{1}{2}$	1

If an event is more likely to happen than another event, then it will be further to the right on the scale. An unlikely event will be somewhere between 0 and 0.5, while a likely event will be between 0.5 and 1. Sometimes a probability scale will have percentages instead of decimals. In this case the 0 will be 0%, 0.5 will be 50% and 1 will be 100%.

Probability of an event not occurring

Knowing that the probability of an event certain to happen is 1 and that the probability of an event impossible to happen is 0, we can calculate the probability of an event happening if we already know the probability of its not happening. We can show this as a simple formula:

Probability of event occurring = 1 − probability of event not occurring

Example 2	From a bag of 8 coloured counters, the probability of *not* selecting a green counter is 0.25. Calculate the probability of selecting a green counter.
	Probability of green = 1 − 0.25 = 0.75

Relative frequency

If we know the theoretical probability of an event happening, then we make a reasonable estimation of what to expect in practice.

For example, if we have a dice numbered 1 to 6, then from 120 throws of that dice we would expect each number to occur 20 times (one sixth of 120).

Example 3	A car manufacturer estimates that the probability of their car developing a fault in the first year is 0.05. If the manufacturer produces 200 000 cars, how many would it expect to develop a fault after one year?
	Expected number of faulty cars = 0.05 × 200 000
	= 10 000

The example below shows how experimental results can be used to estimate relative frequency. The best estimate for relative frequency is obtained from the highest number of trials as any inconsistencies have been minimised.

Example 4	Kiera throws a drawing pin in the air and records the number of times that it lands with its point up. The table below shows her results.

(a) What is the relative frequency of the drawing pin landing point up after 100 throws?
(b) Write down the best estimate for the relative frequency of the drawing pin landing point up from the table. Explain your answer.

Number of trials	10	20	100	200	1000
Point up	8	13	38	70	355

(a) Relative frequency after 100 throws = 38/100 = 0.38
(b) Best estimate for the relative frequency = 355/1000 = 0.355

■ The answer to (b) is obtained from the most trials. This is the most reliable estimate because inconsistent or lucky results will be less likely to affect the total.

34 | Probability

Calculate the probability of simple combined events, using possibility diagrams and tree diagrams where appropriate (in possibility diagrams outcomes will be represented by points on a grid and in tree diagrams outcomes will be written at the end of branches and probabilities by the side of the branches)

Combined events

There are two types of combined event:

- **Independent event**, where the first event has no effect on the second event.
- **Dependent event**, where the first event has an effect on the second event.

When dealing with combined probabilities we can use the **and/or** rule as shown:

> P(1st event) and P(2nd event) = P(1st event) × P(2nd event)
> P(1st event) or P(2nd event) = P(1st event) + P(2nd event)

where P(1st event) is the probability of the first event happening and P(2nd event) is the probability of the second event happening.

If we have to find the probability of the first event happening **and** also the second event happening, we can see from the rule that we **multiply** the probabilities together.

If we have to find the probability of **either** the first event happening **or** the second event happening, we can see from the rule that we **add** the probabilities together. So, remember, in probability:

> **and = multiply (×)   or = add (+)**

In the following example the first event has no effect on the second event, so the events are **independent**.

Example 1 From two throws of a dice find the probability of getting:
  (a) two sixes
  (b) one six followed by an even number (in order)
  (c) a six from either dice

  (a) $P(6)$ and $P(6) = \dfrac{1}{6} \times \dfrac{1}{6} = \dfrac{1}{36}$ For the *and* probabilities we have multiplied and for the *or* probabilities we have added.

  (b) $P(6)$ and $P(\text{even}) = \dfrac{1}{6} \times \dfrac{3}{6} = \dfrac{1}{12}$

  (c) $P(6)$ or $P(6) = \dfrac{1}{6} + \dfrac{1}{6} = \dfrac{1}{3}$ We have also simplified each of the answers, where possible.

In the next example the or/and rules still apply but the first event has an effect on the second event – it is **dependent**.

Example 2 Three red, 10 white, 5 blue and 2 green counters are put into a bag. Out of two selections, without replacing the first selection, find the probability of getting:
  (a) two red counters   (b) a red and a blue counter in that order
  (c) a red and a blue counter in any order   (d) two counters of the same colour.

  (a) $P(\text{red})$ **and** $P(\text{red}) = \dfrac{3}{20} \times \dfrac{2}{19} = \dfrac{3}{190}$

  (b) $P(\text{red})$ **and** $P(\text{blue}) = \dfrac{3}{20} \times \dfrac{5}{19} = \dfrac{3}{76}$

  Simplified answers.

(c) [P(red) **and** P(blue)] **or** [P(blue) **and** P(red)]

$$= \left(\frac{3}{20} \times \frac{5}{19}\right) + \left(\frac{5}{20} \times \frac{3}{19}\right)$$

Note that the order could be red/blue or blue/red, so we have to calculate probabilities of each and add them.

$$= \frac{3}{76} + \frac{3}{76} = \frac{3}{38}$$

(d) If we let red = r, blue = b, green = g, white = w,

[P(r) **and** P(r)] **or** [P(b) **and** P(b)] **or** [P(g) **and** P(g)] **or** [P(w) **and** P(w)]

$$= \left(\frac{3}{20} \times \frac{2}{19}\right) + \left(\frac{5}{20} \times \frac{4}{19}\right) + \left(\frac{2}{20} \times \frac{1}{19}\right) + \left(\frac{10}{20} \times \frac{9}{19}\right)$$

$$= \frac{3}{190} + \frac{1}{19} + \frac{1}{190} + \frac{9}{38}$$

$$= \frac{59}{190}$$ Simplified answers.

■ Note how in each part of the question, the first event had an effect on the second event because there was no replacement of the first selection. For example, when calculating the probabilities of two reds, there was one less red and one less counter altogether available for the second selection.

Possibility diagrams

Possibility diagrams are used for showing all possible results of two combined events. For example, if we throw two dice together and add up the scores each time, we could get the following results:

$$1 + 1 = 2$$
$$1 + 2 = 3$$
$$2 + 1 = 3$$
$$2 + 2 = 4$$
$$\dots = \dots$$

This would obviously take a long time as we have to account for all possibilities. We will set up a possibility diagram instead, where the first event is shown horizontally and the second event is shown vertically.

First dice score

Second dice		1	2	3	4	5	6
	1	2	3	4	5	6	7
	2	3	4	5	6	7	8
	3	4	5	6	7	8	9
	4	5	6	7	8	9	10
	5	6	7	8	9	10	11
	6	7	8	9	10	11	12

The table shows all the possible results from adding the two dice.

We can see that there are 36 different possibilities from the two events.

Example 3   Find the probability of getting a total of 7 from a throw of the two dice.

If we look at the diagram, we can see that the score of seven occurs 6 times out of a total of 36 possible outcomes:

$$P(7) = \frac{6}{36} = \frac{1}{6}$$ Simplified answer.

■ We could have worked this out without using a possibility diagram: 6 ways

$$\begin{cases} 1 + 6 = 7 \\ 6 + 1 = 7 \\ 2 + 5 = 7 \\ 5 + 2 = 7 \\ 3 + 4 = 7 \\ 4 + 3 = 7 \end{cases}$$

Tree diagrams

When two or more combined events are being considered, a **tree diagram** can be used.

A tree diagram displays all the possible outcomes of an event as branches or paths and each branch is labelled as a probability.

For example, if a coin is tossed twice and the results recorded, there are four possibilities: heads and heads, heads and tails, tails and heads or tails and tails. We can show this as a tree diagram, with the combined probabilities.

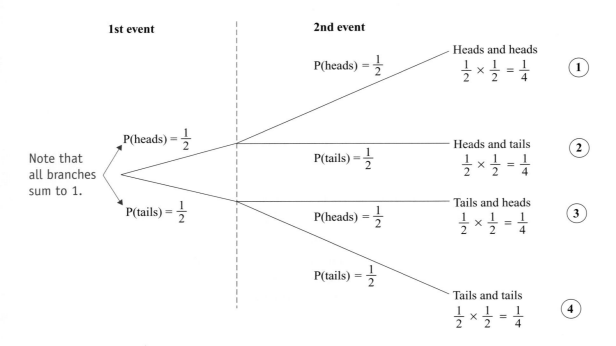

This tree diagram is for a combined independent event as the first coin throw will have no effect on the second event. Note that the branches in each selection add up to 1 because it is absolutely certain that you will get either heads or tails.

If we want to know the probability of, for example, getting either two heads or two tails, we add the appropriate branches (1 and 4):

$\frac{1}{4} + \frac{1}{4} = \frac{1}{2}$ We can say that when you go *along* the branches you *multiply* the probabilities, and when you have *alternative* branches you *add* the total probabilities.

The next example shows a dependent event.

Example 4   A box contains 4 black balls and 6 red balls. A ball is drawn from the box and is not replaced. A second ball is then drawn. Find the probabilities of:
(a) red then black being drawn
(b) red then black in any order
(c) two of the same colour being drawn

As we only have two different colours, we will have two branches for each selection:

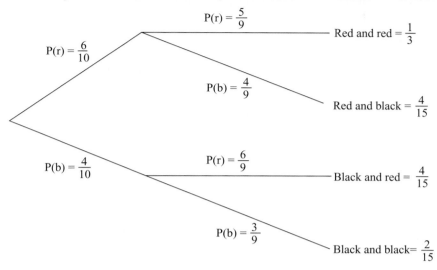

(a) P(red) and P(black) $= \dfrac{6}{10} \times \dfrac{4}{9} = \dfrac{4}{15}$

(b) P(red) and P(black) or P(black) and P(red) $= \dfrac{4}{15} + \dfrac{4}{15} = \dfrac{8}{15}$

(c) P(red) and P(red) or P(black) and P(black) $= \dfrac{1}{3} + \dfrac{2}{15} = \dfrac{7}{15}$

35 | Vectors in two dimensions

Describe a translation by using a vector represented by $\begin{pmatrix} x \\ y \end{pmatrix}$, $\overrightarrow{AB}$ or **a**; add vectors and multiply a vector by a scalar

A vector is a two-dimensional movement given in the form $\begin{pmatrix} x \\ y \end{pmatrix}$ where x is the horizontal movement and y is the vertical movement.

Consider the vector $\overrightarrow{AB} = \begin{pmatrix} 2 \\ 3 \end{pmatrix}$

This vector starts at point A and moves 2 units across and 3 units vertically. This can be shown as a diagram.

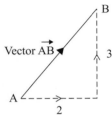

Note that the vector moves to the right which is the positive x-direction and then moves vertically up which is the positive y-direction.

Now consider a vector with negative x- and y-values, for example $\overrightarrow{CD} = \begin{pmatrix} -2 \\ -3 \end{pmatrix}$.

This vector starts at point C, moves 2 units across in the other direction and 3 units down.

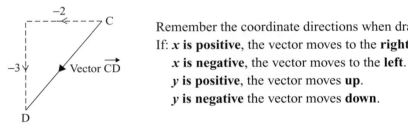

Remember the coordinate directions when drawing these vectors.

If: **x is positive**, the vector moves to the **right**.

    x is negative, the vector moves to the **left**.

    y is positive, the vector moves **up**.

    y is negative the vector moves **down**.

The vector could also be given a letter value, for example

$\mathbf{a} = \begin{pmatrix} 4 \\ -2 \end{pmatrix}$ means that the vector **a** moves 4 units across and 2 units down.

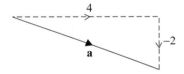

Adding vectors

When adding vectors we simply add the x-values together and the y-values together. For example, if we have two vectors $\overrightarrow{AB}$ and $\overrightarrow{BC}$, where $\overrightarrow{AB} = \begin{pmatrix} 2 \\ -3 \end{pmatrix}$ and $\overrightarrow{BC} = \begin{pmatrix} 3 \\ 1 \end{pmatrix}$,

$$\text{then } \overrightarrow{AB} + \overrightarrow{BC} = \begin{pmatrix} 2+3 \\ -3+1 \end{pmatrix}$$

$$= \begin{pmatrix} 5 \\ -2 \end{pmatrix}$$

The final vector, $\overrightarrow{AC}$ is 5 units across and 2 units down, as this diagram shows.

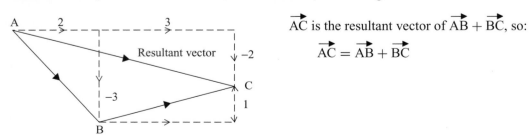

$\overrightarrow{AC}$ is the resultant vector of $\overrightarrow{AB} + \overrightarrow{BC}$, so:

$$\overrightarrow{AC} = \overrightarrow{AB} + \overrightarrow{BC}$$

Multiplying vectors by scalars

A scalar is a numerical value, e.g. 1, 2, 3, −1, −0.5, which, when multiplied by a vector, will change its size. For example, if we have a vector $\overrightarrow{AB} = \begin{pmatrix} 3 \\ -2 \end{pmatrix}$ and we multiply it by the scalar quantity 3, it becomes $3\overrightarrow{AB}$.

Its new size is $3 \times \begin{pmatrix} 3 \\ -2 \end{pmatrix} = \begin{pmatrix} 9 \\ -6 \end{pmatrix}$ We multiply both the x and the y values by 3.

We can show this as a diagram.

The vector $3\overrightarrow{AB}$ is 3 times the size of the vector $\overrightarrow{AB}$.

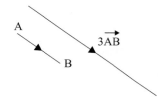

■ Note that the vectors are still in the same direction but the first vector has been increased by a factor of 3 (enlarged by 3).

35 | Vectors in two dimensions

Calculate the magnitude of a vector using Pythagoras' theorem; use modulus signs to denote magnitude, e.g. $|\overrightarrow{AB}|$ or $|\mathbf{a}|$; represent vectors by directed line segments; use the sum and difference of two vectors to express given vectors in terms of two coplanar vectors; use position vectors

The magnitude of a vector |a|

The **magnitude** or **modulus** of a vector is its length.

Example 1 $\overrightarrow{AB} = \begin{pmatrix} 3 \\ 4 \end{pmatrix}$. Find the magnitude of vector $\overrightarrow{AB}$, i.e. $|\overrightarrow{AB}|$.

We first sketch the vector.

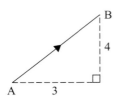

We can see that the x- and y-dimensions form the two shorter sides of a right-angled triangle and AB forms the longer side.

We can then use Pythagoras' theorem to find the longer side of a right-angled triangle:

$|\overrightarrow{AB}| = \sqrt{x^2 + y^2}$ where $\overrightarrow{AB}$ is the longer side
$|\overrightarrow{AB}| = \sqrt{3^2 + 4^2}$ $= 5$ units long

The same method can be used for negative vectors. Try the above example using negative values of x and y. You should get the answer 5 units again.

Representing vectors by line segments

We know that vectors can be shown as letters. We can also show the addition of vectors in terms of letters.

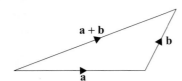

Resultant vector = vector **a** plus vector **b**
$= \mathbf{a} + \mathbf{b}$

■   The resultant vector follows the same general direction, i.e. it starts at the same point as vector **a** and finishes at the end point of vector **b**.

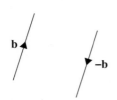

A negative vector can be called an inverse vector. The diagram shows the inverse vector of **b** which is $-\mathbf{b}$. We can see that the vectors are equal in size but opposite in direction.

We can show the subtraction of vectors using letter notation. Here is the resultant vector of $\mathbf{a} - \mathbf{b}$.

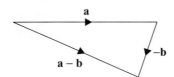

The vector $-\mathbf{b}$ is in the opposite direction to the vector **b** in the diagram above.

The resultant vector $= \mathbf{a} - \mathbf{b} = \mathbf{a} + (-\mathbf{b})$.

This example involves several vectors.

Example 2

In the figure below ABCDEF is a regular hexagon. If $\overrightarrow{AB} = \mathbf{a}$, $\overrightarrow{AF} = \mathbf{b}$ and $\overrightarrow{BC} = \mathbf{c}$, express each of the following in terms of $\mathbf{a}$, $\mathbf{b}$ and $\mathbf{c}$:

(a) $\overrightarrow{DC}$ (b) $\overrightarrow{DE}$ (c) $\overrightarrow{FE}$ (d) $\overrightarrow{FC}$ (e) $\overrightarrow{AE}$

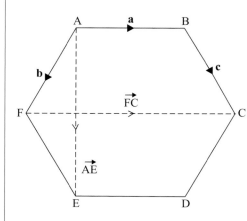

(a) $\overrightarrow{DC} = -\overrightarrow{AF} = -\mathbf{b}$

It is equal in length but in the opposite direction to $\overrightarrow{AF}$.

(b) $\overrightarrow{DE} = -\overrightarrow{AB} = -\mathbf{a}$

It is equal in length but in the opposite direction to $\overrightarrow{AB}$.

(c) $\overrightarrow{FE} = \overrightarrow{BC} = \mathbf{c}$

It is equal in length and in the same direction as $\overrightarrow{BC}$.

(d) $\overrightarrow{FC} = \overrightarrow{FA} + \overrightarrow{AB} + \overrightarrow{BC}$
$= -\mathbf{b} + \mathbf{a} + \mathbf{c}$
$= \mathbf{a} - \mathbf{b} + \mathbf{c}$

See the path of the resultant vector from F to C.

(e) $\overrightarrow{AE} = \overrightarrow{AF} + \overrightarrow{FE} = \mathbf{b} + \mathbf{c}$

See the path of the resultant vector from A to E.

■ There are alternative paths that the resultant vectors could take. For example $\overrightarrow{FC}$ could also take the path $\overrightarrow{FE} + \overrightarrow{ED} + \overrightarrow{DC}$, which still gives the same answer as before: $\mathbf{a} - \mathbf{b} + \mathbf{c}$.

Position vectors

These are vectors which start at the origin O, coordinates (0,0). The method of finding the vectors is the same as before.

Example 3

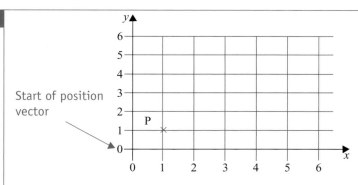

P is the point (1, 1). The vector $\mathbf{a} = \begin{pmatrix} 1 \\ 2 \end{pmatrix}$ and $\mathbf{b} = \begin{pmatrix} 2 \\ -1 \end{pmatrix}$.

(a) Find the vector $\mathbf{a} + 2\mathbf{b}$.

(b) $\overrightarrow{PQ} = \mathbf{a} + 2\mathbf{b}$. Find the position vector of Q.

(c) Calculate $|\mathbf{a}|$, the magnitude of $\mathbf{a}$.

(a) $\mathbf{a} + 2\mathbf{b} = \begin{pmatrix} 1 \\ 2 \end{pmatrix} + 2\begin{pmatrix} 2 \\ -1 \end{pmatrix} = \begin{pmatrix} 5 \\ 0 \end{pmatrix}$

(b) $\overrightarrow{PQ} = \mathbf{a} + 2\mathbf{b}$

The position vector of Q is its vector from the origin (0,0), $\overrightarrow{OQ}$.

$$\overrightarrow{OQ} = \overrightarrow{OP} + \overrightarrow{PQ} = \begin{pmatrix} 1 \\ 1 \end{pmatrix} + \begin{pmatrix} 5 \\ 0 \end{pmatrix} = \begin{pmatrix} 6 \\ 1 \end{pmatrix}.$$

(c) $|\mathbf{a}|$, magnitude of $\mathbf{a} = \sqrt{1^2 + 2^2} = 2.2$ units

■ Note how the method is exactly the same as before, the only difference being that the vector started from the origin and went to another point stated.

36 | Matrices

Understand notation of matrices; add, subtract, multiply matrices; solve equations involving matrices; calculate determinant and inverse of a 2 by 2 matrix

Order of a matrix

The **order of a matrix** gives the number of rows followed by the number of columns in the matrix.

Matrix order = Number of rows × Number of columns

The order of this matrix is therefore 2×2.

We say it is 'of the order 2 by 2'.

These are matrices of different orders:

$$\begin{array}{ccc} & \text{Column 1} & \text{Column 2} \\ \text{Row 1} & \\ \text{Row 2} & \end{array} \begin{pmatrix} a & b \\ c & d \end{pmatrix}$$

$(a \quad b)$

Order 1×2

$\begin{pmatrix} a & b & c \\ d & e & f \end{pmatrix}$

Order 2×3

$\begin{pmatrix} a & b & c \\ c & d & e \\ f & g & h \end{pmatrix}$

Order 3×3

$(a \quad b \quad c)$

Order 1×3

A matrix is often identified by a bold capital letter, for example $\mathbf{A} = \begin{pmatrix} 1 & 2 \\ 3 & 4 \end{pmatrix}$. In a matrix, each number is known as an element, so 1, 2, 3 and 4 are all elements of $\mathbf{A}$.

Addition and subtraction of two or more matrices

The general rule is that only matrices **of the same order** can be added or subtracted.

For example, a 2×2 matrix can be added to another 2×2 matrix and a 1×3 matrix can be added to another 1×3 matrix.

Example 1

If $\mathbf{A} = \begin{pmatrix} 3 & -2 \\ 4 & -3 \end{pmatrix}$ and $\mathbf{B} = \begin{pmatrix} -3 & 6 \\ -4 & 2 \end{pmatrix}$, find (a) $\mathbf{A} + \mathbf{B}$ (b) $\mathbf{A} - \mathbf{B}$

(a) $\begin{pmatrix} 3 & -2 \\ 4 & -3 \end{pmatrix} + \begin{pmatrix} -3 & 6 \\ -4 & 2 \end{pmatrix} = \begin{pmatrix} 3 + (-3) & (-2) + 6 \\ 4 + (-4) & (-3) + 2 \end{pmatrix}$

$= \begin{pmatrix} 0 & 4 \\ 0 & -1 \end{pmatrix}$

Each element in the first matrix is added to the element in the corresponding position in the second matrix, e.g. the number at (row 1, column 1) in matrix $\mathbf{A}$ is added to the number at (row 1, column 1) in matrix $\mathbf{B}$.

(b) $\begin{pmatrix} 3 & -2 \\ 4 & -3 \end{pmatrix} - \begin{pmatrix} -3 & 6 \\ -4 & 2 \end{pmatrix} = \begin{pmatrix} 3 - (-3) & (-2) - 6 \\ 4 - (-4) & (-3) - 2 \end{pmatrix}$

$= \begin{pmatrix} 6 & -8 \\ 8 & -5 \end{pmatrix}$

Take care with double negatives.

■ The same rules apply for all types of matrices whatever the size.

The product of two matrices

These are the rules we need to follow to find the product of two matrices:

- The number of rows in the first matrix must equal the number of columns in the second matrix.
- The order of the resulting matrix is the number of rows in the first matrix multiplied by the number of columns in the second matrix.
- When multiplying, multiply the elements in a row of the first matrix by the elements in a column of the second matrix and add the products. For example, the sum of the products of the elements in the first row of the first matrix and the elements in the second column of the second matrix gives the element in the first row and second column of the product matrix.

When you multiply two matrices together, the final matrix will take this form:

$$\begin{pmatrix} \text{row } 1 \times \text{column } 1 & \text{row } 1 \times \text{column } 2 \\ \text{row } 2 \times \text{column } 1 & \text{row } 2 \times \text{column } 2 \end{pmatrix}$$

Example 2

If $\mathbf{A} = \begin{pmatrix} 3 & -2 \\ 4 & -3 \end{pmatrix}$ and $\mathbf{B} = \begin{pmatrix} -3 & 6 \\ -4 & 2 \end{pmatrix}$, find $\mathbf{A} \times \mathbf{B}$.

$$\begin{pmatrix} 3 & -2 \\ 4 & -3 \end{pmatrix} \times \begin{pmatrix} -3 & 6 \\ -4 & 2 \end{pmatrix} = \begin{pmatrix} 3 \times (-3) + (-2) \times (-4) & 3 \times 6 + (-2) \times 2 \\ 4 \times (-3) + (-3) \times (-4) & 4 \times 6 + (-3) \times 2 \end{pmatrix}$$

$$= \begin{pmatrix} -9 + 8 & 18 + (-4) \\ -12 + 12 & 24 + (-6) \end{pmatrix} = \begin{pmatrix} -1 & 14 \\ 0 & 18 \end{pmatrix}$$

■ Note that matrix multiplication is not commutative, so $\mathbf{B} \times \mathbf{A}$ would give $\begin{pmatrix} 15 & -12 \\ -4 & 2 \end{pmatrix}$.

Scalar multiplication of a matrix

When you multiply a matrix by a scalar quantity, every element of the matrix is multiplied by that number. For example, if the scalar quantity is 2 then all elements in the matrix are multiplied by 2.

Example 3

If $\mathbf{C} = \begin{pmatrix} 2 & -1 \\ 4 & -3 \end{pmatrix}$ and $\mathbf{D} = \begin{pmatrix} 4 & -3 \\ -1 & 2 \end{pmatrix}$, find (a) $2\mathbf{C}$ (b) $-3\mathbf{D}$ (c) $2\mathbf{C} - 3\mathbf{D}$.

(a) $2\mathbf{C} = 2\begin{pmatrix} 2 & -1 \\ 4 & -3 \end{pmatrix} = \begin{pmatrix} 4 & -2 \\ 8 & -6 \end{pmatrix}$

Every term inside the matrix has been multiplied by the scalar quantity of 2.

(b) $-3\mathbf{D} = -3\begin{pmatrix} 4 & -3 \\ -1 & 2 \end{pmatrix} = \begin{pmatrix} -12 & 9 \\ 3 & -6 \end{pmatrix}$

Every term inside the matrix has been multiplied by the scalar quantity of -3.

(c) $2\mathbf{C} - 3\mathbf{D} = \begin{pmatrix} 4 & -2 \\ 8 & -6 \end{pmatrix} + \begin{pmatrix} -12 & 9 \\ 3 & -6 \end{pmatrix} = \begin{pmatrix} -8 & 7 \\ 11 & -12 \end{pmatrix}$

■ The scalar quantity can be positive, negative or a fraction: every element inside the matrix is multiplied by the scalar quantity.

We have only looked at 2×2 matrices so far with scalar quantities, but the method is essentially the same for matrices of other orders.

Using algebra within matrices

Example 4

If $\begin{pmatrix} 2 & x \\ 4 & y \end{pmatrix} \begin{pmatrix} 5 \\ 2 \end{pmatrix} = \begin{pmatrix} 14 \\ 30 \end{pmatrix}$, find x and y.

First find the left-hand side of the problem by multiplying the matrices together:

$$\begin{pmatrix} 2 & x \\ 4 & y \end{pmatrix} \begin{pmatrix} 5 \\ 2 \end{pmatrix} = \begin{pmatrix} 2 \times 5 + x \times 2 \\ 4 \times 5 + y \times 2 \end{pmatrix} = \begin{pmatrix} 10 + 2x \\ 20 + 2y \end{pmatrix}$$

Now we can rewrite the problem:

$$\begin{pmatrix} 10 + 2x \\ 20 + 2y \end{pmatrix} = \begin{pmatrix} 14 \\ 30 \end{pmatrix}$$

As the left-hand side equals the right-hand side, we can remove the matrices and now treat the problem as two linear equations.

So $\begin{aligned} 10 + 2x &= 14 \\ 2x &= 14 - 10 \\ 2x &= 4 \\ x &= 2 \end{aligned}$ and $\begin{aligned} 20 + 2y &= 30 \\ 2y &= 30 - 20 \\ 2y &= 10 \\ y &= 5 \end{aligned}$

■ Note that we can use this method on other orders of matrices, provided that it is *possible* to multiply the matrices together.

Zero matrix

The **zero matrix** is ones in which all the elements are zero. It is sometimes called a **null** matrix.

This is a 2×2 null matrix: $\begin{pmatrix} 0 & 0 \\ 0 & 0 \end{pmatrix}$

The identity matrix (I)

A **diagonal matrix** has all its elements zero except for those in the leading diagonal (from top left to bottom right).

The **identity matrix**, **I**, is a diagonal matrix with the elements equal to 1.

This is the 2×2 identity matrix: $\begin{pmatrix} 1 & 0 \\ 0 & 1 \end{pmatrix}$

If you multiply a matrix by the identity matrix the result is the original matrix.

So, if $\mathbf{A} = \begin{pmatrix} 2 & 3 \\ 1 & 4 \end{pmatrix}$ then $\mathbf{A} \times \mathbf{I} = \begin{pmatrix} 2 & 3 \\ 1 & 4 \end{pmatrix}$.

Calculating the determinant of a matrix |A|

If $\mathbf{A} = \begin{pmatrix} a & b \\ c & d \end{pmatrix}$, then the determinant of **A**, given the symbol |**A**|, is found using the following formula:

$$|\mathbf{A}| = ad - bc$$

where the values of a, b, c and d are taken from their relative positions in the 2×2 matrix.

Example 5

Find the determinant of the following matrix: $\mathbf{X} = \begin{pmatrix} 3 & 5 \\ 2 & 4 \end{pmatrix}$.

Using the formula above, $\begin{aligned} |\mathbf{A}| &= ad - bc \\ |\mathbf{X}| &= 3 \times 4 - 5 \times 2 \\ |\mathbf{X}| &= 2 \end{aligned}$

Note that the determinant can also be a negative value, depending on the value of the elements inside the matrix.

Inverse of a matrix (A^{-1})

As its name suggests, this is simply the inverse or reciprocal of a matrix. If we multiply any matrix by its inverse we will obtain the identity matrix:

$$AA^{-1} = I$$

To find the inverse of the 2×2 matrix, $\mathbf{A} = \begin{pmatrix} a & b \\ c & d \end{pmatrix}$, use the formula:

$$\mathbf{A}^{-1} = \frac{1}{|\mathbf{A}|} \begin{pmatrix} d & -b \\ -c & a \end{pmatrix}$$

Note how we have interchanged the a and d values and changed the signs on the b and c values.

Example 6

Find the inverse of the following matrix and perform the appropriate check to ensure that is correct: $\quad \mathbf{A} = \begin{pmatrix} 4 & 1 \\ 2 & 3 \end{pmatrix}$

Using the formula for the inverse,

$$\mathbf{A}^{-1} = \frac{1}{(4 \times 3 - 2 \times 1)} \begin{pmatrix} 3 & -1 \\ -2 & 4 \end{pmatrix}$$

$$= \frac{1}{10} \begin{pmatrix} 3 & -1 \\ -2 & 4 \end{pmatrix}$$

$$= \begin{pmatrix} 0.3 & -0.1 \\ -0.2 & 0.4 \end{pmatrix}$$

To check, we need to show that $\mathbf{AA}^{-1} = \mathbf{I}$:

$$\begin{pmatrix} 4 & 1 \\ 2 & 3 \end{pmatrix} \begin{pmatrix} 0.3 & -0.1 \\ -0.2 & 0.4 \end{pmatrix} = \begin{pmatrix} 4 \times 0.3 + 1 \times (-0.2) & 4 \times (-0.1) + 1 \times 0.4 \\ 2 \times 0.3 + 3 \times (-0.2) & 2 \times (-0.1) + 3 \times 0.4 \end{pmatrix}$$

$$= \begin{pmatrix} 1 & 0 \\ 0 & 1 \end{pmatrix}$$

---

37 | Transformations

Use transformations to describe and perform reflections, rotations (multiples of 90°), translations, enlargements

An **object** undergoing a transformation changes either in position or in shape. The object's new position or shape is called its **image**.

There are four types of transformation:

- reflection
- rotation
- translation
- enlargement

Reflection

When an object is **reflected** it undergoes a 'flip' movement about a mirror line. The size of the object remains unchanged. Here are some examples of reflections:

Reflection of shape in mirror line

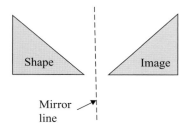

Reflection of shape in x-axis and y-axis

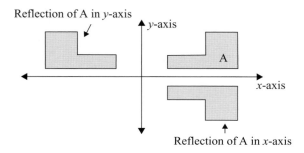

Rotation

When an object is **rotated**, it undergoes a turning movement about a specific point called the **centre of rotation**. When we describe a rotation we need to state:

- the position of the centre of rotation
- the direction of rotation
- the angle of rotation

Here are some examples of rotations:

A 90° clockwise rotation

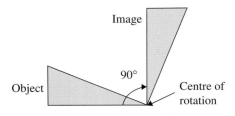

90° rotations using the origin as centre of rotation

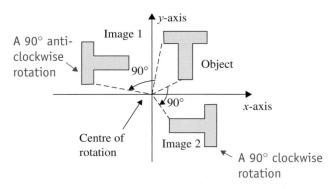

A 90° clockwise rotation using the mid-point of an edge as the centre of rotation

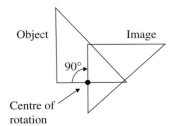

Object   Image

90°

Centre of rotation

Each corner of the triangle is moved 90° using the centre of rotation as the reference point for the protractor. The distance from the centre of rotation to the same corners on the object and the image must be the same.

When performing rotations, use a protractor to give accurate angles of rotation and ensure that the distances of the same points on the object and image are the same. (Use compasses to help with this.)

Translation

When an object is **translated**, it undergoes a vector type movement, i.e. the whole shape moves to another location on a grid without changing size or rotating.

A translation of $\begin{pmatrix} 5 \\ 4 \end{pmatrix}$

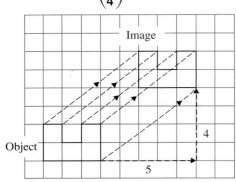

A translation of $\begin{pmatrix} -6 \\ -1 \end{pmatrix}$

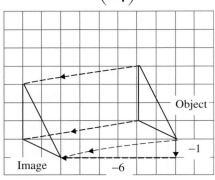

The object has moved 5 units to the right and then 4 units upwards.

The object has moved 6 units to the left and then 1 unit down.

Enlargement

When an object undergoes an **enlargement**, the image is of a similar shape but a different size. When we perform an enlargement, we need to state:

- the position of the centre of enlargement
- the scale factor of enlargement

If we use a scale factor of 2, then the image will be twice the size of the original object. However, if the scale factor of enlargement is less than 1, then the object actually reduces in size.

An enlargement of scale factor 2

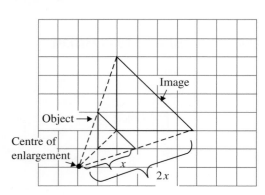

An enlargement of scale factor $\frac{1}{2}$

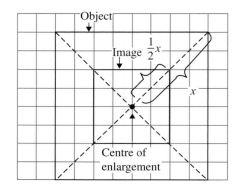

The distance from the centre of enlargement to each vertex is measured with a ruler. This distance is multiplied by the scale factor to give the position of the same vertex on the image.

The distance from the centre of enlargement to each vertex is measured, and then multiplied by the scale factor to give the new position of that vertex of the image.

A negative enlargement of scale factor −2

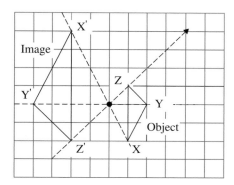

The distance from the centre of enlargement to each vertex is measured with a ruler. This distance is multiplied by the negative scale factor to give the position of the same vertex on the image *but in the opposite direction.*

Note how this gives a similar result to a 180° rotation. In fact, a scale factor of −1 enlargement is exactly the same as a 180° rotation.

Summary

- **Reflection** — The shape reflects about a mirror line to become its image. It remains the same size.
- **Rotation** — The shape rotates about a centre of rotation. The direction of rotation must be stated. The shape remains the same size.
- **Translation** — The shape performs a vector movement on a grid without rotating or changing size.
- **Enlargement** — The whole shape changes in size according to a given scale factor. The centre of enlargement must be stated. A scale factor enlargement greater than 1 will produce a bigger image than the object; a scale factor less than 1 will produce a smaller image than the object.

37 | Transformations

Perform reflections, rotations, translations, enlargements, shears, stretches and their combinations; identify and give precise descriptions of transformations connecting given figures; describe transformations using coordinates and matrices

The fundamentals of reflection, rotation, translation and enlargement have already been covered, but now we need to extend these basic principles.

Reflections – stating the equation of the mirror line

We need to state the exact position of the mirror line on the grid.

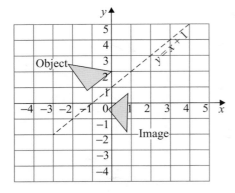

The image has been reflected in the mirror line shown.

We can work out the equation of the mirror line by using the $y = mx + c$ rule covered in Unit 19.

The line intercepts the y-axis at 1 (therefore $c = 1$) and the gradient is also 1 (therefore $m = 1$).

The object has undergone a reflection in the line $y = x + 1$.

Rotations – finding the centre and angle of rotation

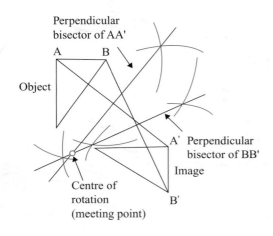

To find the centre of rotation:

1 Join a point on the object to its corresponding point on the image, e.g. join A to A'.
2 Construct the perpendicular bisector of this line.
3 Repeat this process for another point, e.g. join B to B'.
4 The two perpendicular bisectors meet at the centre of rotation.

To find the angle, simply measure with a protractor the angle turned from any point to its corresponding point on the image.

Enlargement – finding the centre and scale factor of enlargement

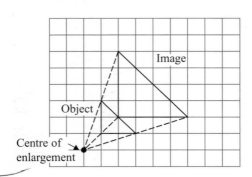

To find the centre of enlargement:

1 Join up any point on the image to its corresponding point on the object. Continue this line back for a few centimetres.
2 Repeat this process for another point.
3 The two lines meet at the centre of enlargement.

To find the scale factor, divide the length of any side of the image by the length of the corresponding side on the object, e.g. $4 \div 2 = 2$.

Shear and stretch transformations

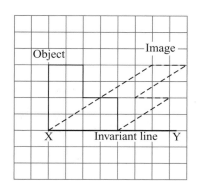

Shear transformation with XY as the invariant line

A **shear transformation** refers to the sliding of different parts of the shape in relation to an **invariant line**.

In the diagram on the left, XY is the invariant line, i.e. it is a fixed point on the diagram.

When describing a shear transformation, we need to state the position of the invariant line and the **shear factor**:

$$\text{Shear factor} = \frac{\text{Distance a point moves due to shear}}{\text{Perpendicular distance of point from invariant line}}$$

In the diagram, the shear factor $= \frac{6}{4} = 1.5$.

Stretch transformation with XY as the invariant line

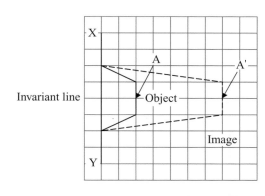

A **stretch transformation** refers to a lengthening of the object in one direction only, in relation to an **invariant line**.

In the diagram on the left, XY is the invariant line, which is a fixed point on the diagram.

When describing a stretch transformation, we need to state the position of the invariant line and the **scale factor**:

$$\text{Scale factor} = \frac{\text{Perpendicular distance of the image of a point from the invariant line}}{\text{Perpendicular distance of the point from the invariant line}}$$

In the diagram, the scale factor $= \frac{7}{2} = 3.5$.

Combinations of transformations

Sometimes we have more than one transformation of a shape on the same grid.

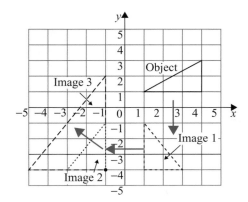

The diagram on the left shows three transformations:

- Object to Image 1 is a rotation of 90° clockwise using the origin as the centre of rotation.
- Image 1 to Image 2 is a reflection in the y-axis.
- Image 2 to Image 3 is an enlargement of scale factor 2, centre of enlargement $(-1, -4)$.

Transformations and matrices

Any shape on a grid can be shown as a matrix by using the coordinate positions of each of its corners (vertices).

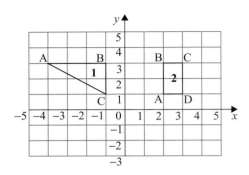

Shape 1 has the matrix:
$$\begin{array}{ccc} A & B & C \end{array}$$
$$\begin{pmatrix} -4 & -1 & -1 \\ 3 & 3 & 1 \end{pmatrix}.$$

Shape 2 is represented by:
$$\begin{array}{cccc} A & B & C & D \end{array}$$
$$\begin{pmatrix} 2 & 2 & 3 & 3 \\ 1 & 3 & 3 & 1 \end{pmatrix}$$

where the first row shows the x-coordinate, and the second row the y-coordinate.

Performing a matrix transformation on a shape

When we multiply the matrix of vertices of a shape by another matrix, we perform a transformation on that shape. We have already studied multiplication of matrices in Unit 36, so we know the method involved.

Example 1

Triangle XYZ has coordinates X $(-4, 4)$, Y $(-4, 1)$ and Z $(-1, 1)$. Draw the image of triangle XYZ under the transformation matrix $\begin{pmatrix} 0 & 1 \\ -1 & 0 \end{pmatrix}$.

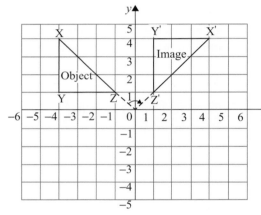

Plot the triangle XYZ using the coordinates given and the calculate the new position of each vertex after the matrix multiplication:

$$\begin{pmatrix} 0 & 1 \\ -1 & 0 \end{pmatrix} \times \begin{array}{c} \begin{array}{ccc} X & Y & Z \end{array} \\ \begin{pmatrix} -4 & -4 & -1 \\ 4 & 1 & 1 \end{pmatrix} \end{array}$$

$$= \begin{array}{c} \begin{array}{ccc} X' & Y' & Z' \end{array} \\ \begin{pmatrix} 4 & 1 & 1 \\ 4 & 4 & 1 \end{pmatrix} \end{array}$$

We then plot these new coordinates on the grid.

We can see that the object has undergone a rotation of $90°$ clockwise with the origin as the centre of rotation.

Transformations involving more than one matrix: A′ = MR(A)

In this case the object or point **A** is first multiplied by the matrix **R** and then the result is multiplied by matrix **M** to give its final position **A′**

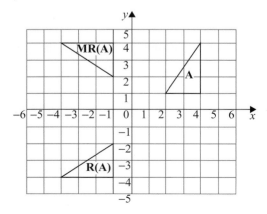

Shape $\mathbf{A} = \begin{pmatrix} 2 & 4 & 4 \\ 1 & 1 & 4 \end{pmatrix}$

Matrix $\mathbf{R} = \begin{pmatrix} 0 & -1 \\ -1 & 0 \end{pmatrix}$

Matrix $\mathbf{M} = \begin{pmatrix} 1 & 0 \\ 0 & 1 \end{pmatrix}$

Matrix **R** performs a reflection in the line $y = -x$ to produce **R(A)** and then matrix **M** performs a reflection in the x-axis to give **MR(A)**.

The combined transformation is therefore a 90° anti-clockwise rotation from the origin.

■ Note that a quicker method would be to multiply **M** by **R** to find the resultant matrix, and then multiply this by **A** to get **A′**. Check for yourself.

Inverse transformations

We may need to find the **inverse** of a transformation, i.e. the operation which reverses the image back to its original position.

- The inverse of an **image reflection** is simply the object itself. (A mirror image is an inverse of a shape.)
- The inverse operation of a **translation** is the inverse of the x- and y-movements, e.g. the inverse of the translation $\begin{pmatrix} 2 \\ -3 \end{pmatrix}$ is the translation $\begin{pmatrix} -2 \\ 3 \end{pmatrix}$.
- The inverse operation of an **enlargement** is to divide the image by the scale factor instead of multiplying.
- The inverse operation of a **rotation** is to rotate the image in the opposite direction using the same angle.

Transformations and inverse matrices

To find the inverse of a matrix transformation, multiply the transformation matrix **A** by the inverse matrix **A**$^{-1}$. (See Unit 36 for revision on inverse matrices.)

So, using triangle XYZ from Example 1, if **A** is the transformation matrix $\begin{pmatrix} 0 & 1 \\ -1 & 0 \end{pmatrix}$, then

$$\mathbf{A}^{-1} \begin{pmatrix} 0 & -1 \\ 1 & 0 \end{pmatrix}.$$

Now, if we multiply the inverse matrix by the image we should get back to the original object:

$$\begin{pmatrix} 0 & -1 \\ 1 & 0 \end{pmatrix} \times \begin{pmatrix} 4 & 1 & 1 \\ 4 & 4 & 1 \end{pmatrix} = \begin{pmatrix} -4 & -4 & -1 \\ 4 & 1 & 1 \end{pmatrix}$$

These are the coordinates of the original trangle XYZ given by a 90° rotation of triangle X′Y′Z′ anti-clockwise, using the origin as the centre of rotation.

Practice exams

These exams contain Paper 1 and Paper 3 type questions. The marks for each part of the question are given on the right-hand side.

Practice core exam 1

1 Given that y is a positive integer, if $y \leq 7$ and $y \neq 4$, list the possible values of y. **(2)**

2 A building labourer is paid $480 each week.
   a One week, the ratio of the money he spends to the money he saves is $7 : 5$. Calculate how much money he saves in that week. **(2)**
   b He spends 20% of the $480 on clothes. Calculate how much he spends on clothes. **(2)**

3 An international athlete starts a marathon at 9.45 a.m. and finishes at 12.03 p.m. How long does he take? Give your answer in hours and minutes. **(2)**

4 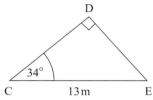 The diagram shows the cross-section of the roof of a garden shed. It is 13 metres long, angle DCE $= 34°$ and angle CDE $= 90°$.

Calculate the lengths of the two sides of the roof, CD and DE. **(4)**

5 The width of a bridge is estimated as 1000 metres. The estimate is correct to the nearest hundred metres. Between what limits does the actual width of the bridge lie? **(2)**

6 Three adults and one child have a meal at a restaurant. The adults' meals all cost the same, the child's meal is half the price. The total bill is $49.70.
What is the cost of one adult's meal? **(2)**

7 a Complete the statement below by putting the correct symbol, $>$ or $<$, between the two numbers.

   $$24.8 \times 10^3 \quad 2.5 \times 10^5$$ **(1)**

   b Which one of the above two numbers is written in standard form? **(1)**
   c Rewrite the other number in standard form. **(1)**

8 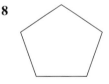 **a** What is the order of rotation symmetry of a regular pentagon, as shown in the diagram? **(1)**
   b How many lines of symmetry does a regular octagon have? **(1)**

9 John has two circular birthmarks on his arm. One has a radius of 1 cm while the other has a radius of 0.7 cm. Calculate the total area of the birthmarks. **(3)**

10 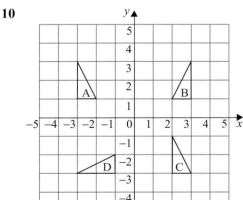 **a** Describe **fully** the single transformation which maps:
   i triangle A onto B **(2)**
   ii triangle A onto C **(2)**
   iii triangle A onto D **(2)**
   b On the grid draw the image of an enlargement of scale factor 2 on triangle B, from centre of enlargement (4, 1). Label your new image E. **(3)**

11

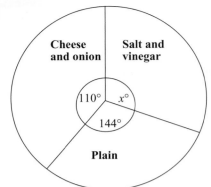

Cheese and onion

Salt and vinegar

110° $x°$

144°

Plain

Five hundred schoolchildren were asked for their favourite flavour crisps and the results are displayed on the pie chart shown.

a What angle should represent Salt and vinegar? **(1)**

b How many schoolchildren said Plain? **(2)**

12 A football shirt costs x and a rugby shirt costs y.

    a **i** Find an expression, in terms of x and y, for the total cost of 4 football shirts and 3 rugby shirts. **(1)**

        ii The total cost of 4 football shirts and 3 rugby shirts is $80. Write down an equation, in terms of x and y, to show this information. **(1)**

    b The total cost of 6 football shirts and 9 rugby shirts is $174. Write down an equation, in terms of x and y, to show this information. **(1)**

    c Solve the simultaneous equations from part **a ii** and part **b** to find the cost of:

      i one football shirt

      ii one rugby shirt. **(4)**

Practice core exam 2

1 Subtract 8.25 from 9.

    Give your answer: **a** as a decimal **(1)**

                    b as a fraction in its lowest terms. **(1)**

2 Find $\sqrt{6.25}$. **(1)**

3 Solve the equation $3y - 4 = y + 6$. **(2)**

4 How many integers between 1 and 75 contain the digit 7? **(2)**

5 In triangle XYZ, angle Y $= 90°$, side XY $= 7$ cm and side XZ $= 8$ cm.

    Calculate: **a** side YZ **(2)**

               b angle Z. **(2)**

6 Santy bought a stereo player for $76.

    After six months she sold it to her friend Yolander making a 20% loss.

    a How much did Yolander pay for it? **(2)**

    b If Yolander later sold it to her friend Maria, making a 20% profit, how much did Maria pay for it? **(2)**

7 Fahad and Madiha share $350 in the ratio 3 : 4. How much does each receive? **(2)**

8   **a** Write $4^0 + 4^2$ as a single number without indices. **(1)**

    b Write 4^{-1} as a decimal. **(1)**

9 Make x the subject of the formula $y = ax + c$. **(2)**

10 Factorise: **a** $5xy + 10x$ **(1)**

             b $x^2 - 4x$ **(1)**

11 A piece of cork is suspended by a length of string from a fixed point above it.

    In the wind, it swings from side to side but the length of the string remains constant.

    Describe or draw the locus of the cork. **(2)**

12 **a** Complete the table for the function $y = x^2 - 4x + 2$ for $-2 \le x \le 5$. **(2)**

x	-2	-1	0	1	2	3	4	5
y	14		2	-1	-2	-1		7

b On a sheet of graph paper, draw the graph of $y = x^2 - 4x + 2$. **(5)**
c Use your graph to estimate:
   i the solutions of the equation $x^2 - 4x + 2 = 0$ **(2)**
   ii the minimum value of $x^2 - 4x + 2$. **(1)**
d Complete the table for the function $y = 5 - x$. **(2)**

x	-2	0	5
y		5	

e On the same grid, draw the graph of $y = 5 - x$ for $-2 \le x \le 5$. **(2)**
f The two graphs intersect at two points A and B. Write down the coordinates of both points of intersection. **(2)**

13 In a recent mathematics test involving 50 students, scored out of a maximum of 10, the following results were obtained:

1	9	2	5	6	5	8	7	6	7	8	3
9	5	4	10	8	7	5	6	6	7	5	2
5	2	6	9	5	7	6	5	6	2	8	6
7	3	3	8	7	6	5	5	6	4	3	4
5	10										

a Use the scores to complete the tally chart below. **(3)**

Score	Tally	Frequency
1	/	1
2		4
3		
4		
5		
6		
7		
8		
9		
10		
		Total = 50

b Calculate the mean score. **(2)**
c Find the median score. **(2)**
d Write down the mode and the range of the scores. **(2)**
e On a piece of graph paper, draw a bar chart to show the information clearly. **(4)**

Practice core exam 3

1 Calculate: **a** $8 + 10 \times 3$ **(1)**

              b $\frac{1}{4}(9 \times 6 - 6 \div 3)$. **(2)**

2 Last year, Ahmed's wages were \$80 per week. His wages are now \$92 per week. Calculate the percentage increase. **(2)**

3 Write 0.455, $\sqrt{0.25}$, 46%, $\frac{9}{20}$ in order of size, using the appropriate inequality signs. **(2)**

4 Bradley has three counters, two of them red and one black. He places them side by side, in random order, on a table. One possible arrangement is:

     black   red   red

    a Write down the other possible arrangements. **(1)**

    b What is the probability that the two red counters are next to each other? **(1)**

5 The daily maximum temperatures in Kuwait City for a week in July were:

     $48\,°C$, $51\,°C$, $49\,°C$, $50\,°C$, $51\,°C$, $50\,°C$, $49\,°C$

    a Find the median temperature. **(2)**

    b Calculate the mean (average) temperature. **(2)**

6 Give an example of:

    a a prime number between 80 and 90 **(1)**

    b an irrational number **(1)**

    c a negative number which is not an integer. **(1)**

7 The vector $\mathbf{a} = \begin{pmatrix} 3 \\ 1 \end{pmatrix}$ and $\mathbf{b} = \begin{pmatrix} -2 \\ -3 \end{pmatrix}$. Find:

    a $\mathbf{a} + \mathbf{b}$ **(1)**

    b $3\mathbf{a} - \mathbf{b}$. **(2)**

8 Solve the simultaneous equations $3a - b = 11$

                          $4a + b = 10$. **(3)**

9

    a Write down the correct mathematical name for the shape on the left. **(1)**

    b Calculate its total surface area. **(2)**

    c Find its volume, giving your answer in mm³. **(2)**

10 £1 (pound) = €1.53 (euros)

    a François changed 10 000 euros into pounds sterling. How many pounds did he receive? **(2)**

    b François took his pounds to London and spent £4000 over a period of two months. He then converted his remaining pounds back into euros, at the same exchange rate. How many euros did he have left? **(2)**

11

    Triangle ABC is right-angled with side AB = 6 cm and side AC = 8 cm. Calculate:

    a the length of side BC **(2)**

    b angle C. **(2)**

12 What is the difference in height between a point 220 metres above sea level and another point 425 metres below sea level? **(1)**

13

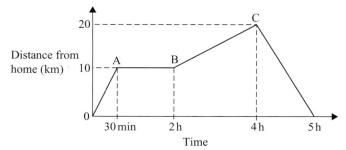

The travel graph shows a cyclist's journey from home to point A, resting for a short period before travelling on to point C, then returning home again.

  a For how many minutes does he rest?   **(1)**

  b At what speed does he travel on the return journey?   **(1)**

  c What is his overall average speed?   **(2)**

14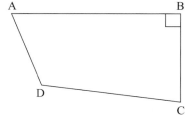

$AB = 12$ cm, $AD = 6$ cm, $DC = 10$ cm, $BC = 8$ cm and angle ABC is $90°$.

  a Copy the diagram on the left to scale.   **(2)**

  b Draw the locus of points which are equidistant from points A and B.   **(2)**

  c Draw the locus of points which are equidistant from lines AB and BC.   **(2)**

  d Label the intersection of the two loci X. Measure the distance DX.   **(2)**

15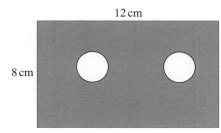

The diagram shows a steel sheet which has had two identical circular holes of radius r cm cut out of it.

  a Explain why the shaded area is $96 - 2\pi r^2$.   **(2)**

  b When the radius $= 1$ cm, calculate:

    i the area of the shaded part   **(2)**

    ii the shaded area as a percentage of the area of the whole rectangle.   **(2)**

  c Calculate the value of r when the shaded area is 90 cm^2.   **(2)**

Practice core exam 4

1 An irregular pentagon has 4 equal angles of $100°$. Calculate the remaining angle.   **(2)**

2 In Saudi Arabia, a can of coke costs 1 Saudi riyal. If £1 = 5.8 SR, how many cans of coke can be bought for the equivalent of £3.20?   **(2)**

3 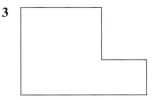 The shape on the left is an irregular hexagon and all the edges are either vertical or horizontal.

What is the sum of the interior angles of the shape?   **(2)**

4 a Write down the next two terms in each of the given sequences.

    i 3, 6, 9, 12, . . .   **(1)**

    ii 24, 12, 6, 3, . . .   **(1)**

    iii 2, 3, 4, 9, 16, 29, 54, . . .   **(2)**

  b The nth term of a sequence is given by $n^2 + 4$.

    i Find the 30th term   **(1)**

    ii Work out which term is equal to 125.   **(2)**

c Consider the sequence: $\frac{2}{5}, \frac{3}{6}, \frac{4}{7}, \frac{5}{8} \cdots$

  i Write down the next term. **(1)**

  ii Write down the 16th term. **(1)**

  iii Work out which term in the sequence is equal to $\frac{7}{8}$. **(1)**

  iv Write down the nth term. **(1)**

5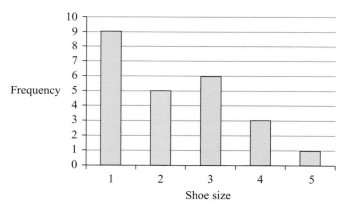

The two parallelograms are similar. Their base measurements are 8 cm and 4 cm respectively. The area of parallelogram A is 40 cm². Calculate:

a the perpendicular height of A **(1)**

b the area of parallelogram B. **(2)**

6 a Simplify $3(x - 4) - 2(x + 3)$. **(2)**

  b List the integer values of y for which $-2 < y \leq 4$. **(2)**

  c Evaluate $4x^3 + 8x^2$

  i when $x = 2$ **(1)**

  ii when $x = -3$. **(2)**

  d Solve $5(x - 4) = 3(x + 2)$. **(2)**

7

In the circle, O is the centre, line DOX is the diameter, line ACY is a tangent to the circle and angle CDX = 42°. Write down the values of:

a angle OCY **(1)**

b angle CXD **(1)**

c angle DCX. **(1)**

8

Shoe sizes of students

The bar chart shows the shoe sizes of several children aged between 8 and 9 years old.

a How many children took part in the survey? **(2)**

b Write down the mode. **(1)**

c Write down the median. **(2)**

d Write down the mean (average). **(2)**

9 a Calculate the volume of a cube of side 4 cm. **(1)**

  b A box, in the shape of a cuboid, is 28 cm long, 24 cm wide and 44 cm deep. How many cubes of side 4 cm can fit into it? **(2)**

10 $\overrightarrow{AB} = \begin{pmatrix} 5 \\ -2 \end{pmatrix}$ and $\overrightarrow{BC} = \begin{pmatrix} 1 \\ -4 \end{pmatrix}$

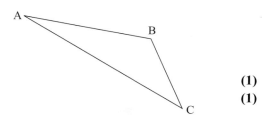

Write down, as a column vector:

a $\overrightarrow{AC}$ (1)

b $3\overrightarrow{AC}$. (1)

11

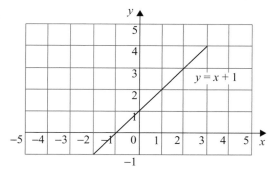

The graph of $y = x + 1$ is drawn on the grid above.

a On a copy of the grid, draw the graph of $y = 5 - x$. (2)

b Use your graphs to solve the simultaneous equations: $y - x = 1$ (2)
$$y + x = 5.$$ (2)

12 A map has a scale of 1 : 30 000.

a A road on the map measures 3 cm long.
What is the real length of the road in kilometres? (2)

b Another road is actually 3.0 km. What would its length be on the map? (2)

13 Describe fully the symmetry of an isosceles trapezium. (2)

14

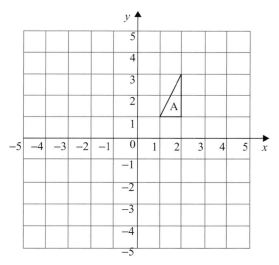

Copy the grid and perform the following transformations on A:

a Reflection in the x-axis.
Label the image B. (2)

b Rotation by 180°, using the origin as centre of rotation.
Label the image C. (2)

c Enlargement of scale factor 0.5 from coordinate (4, 1).
Label the image D. (2)

15 The population of a town increases by 5% each year. If in 2002 the population was 85 000, in which year is the population expected to exceed 100 000 for the first time? (3)

Practice core exam 5

1 $x = \dfrac{3a - c^2}{de}$. Make a the subject of the formula. (2)

2 Calculate: **a** 3^{-4} (1)
b y^0 (1)
c $4^2 \div 4^{-3}$. (1)

3 Factorise: **a** $2x^2 - 8xy^2$ (1)
b $28a^3b^2c + 7a^4b^3c^2 - 14a^2b^2c^3$. (2)

4 Solve the equation $3y + 17 = 21 - 5y$. (2)

5 A school measures the dimensions of its rectangular playing field to the nearest metre. The length is recorded as 250 m and its width as 165 m. Express the range in which the length and width lie using inequalities. (2)

6 The owner of a shop bought 230 shirts for $2472.50.
  a How much did she pay for each shirt? (1)
  b She put the shirts on sale at $12.90 each. Calculate the percentage profit on each. (2)

7 a Complete the table for the function $y = 4 - 2x - x^2$ for $-4 \le x \le 2$. (3)

x	-4	-3.5	-3	-2	-1	0	1	1.5	2
y		-1.25	1	4		4	1		-4

  b On a sheet of graph paper, draw the graph of $y = 4 - 2x - x^2$. (5)
  c Use your graph to estimate:
    i the solutions of the equation $4 - 2x - x^2 = 0$ (2)
    ii the maximum value of $4 - 2x - x^2$. (1)
  d On the same grid, draw the graph of $y = x$. (2)
  e Use the two graphs to solve the equation $4 - 2x - x^2 = x$. (2)

8 a The factors of 24 are 1, 2, 3, 4, 6, 8, 12 and 24. List all the factors of 36. (1)
  b The prime factors of 24 are $2 \times 2 \times 2 \times 3 = 2^3 \times 3$.
    Write in a similar way, the prime factors of 36. (1)
  c Find the highest common factor of 24 and 36. (2)
  d Find the lowest common multiple of 24 and 36. (2)
  e Find the lowest common multiple of 24 000 and 36 000. (1)

9 The ratio of the interior angles of a pentagon is $2 : 3 : 4 : 4 : 5$.
Calculate the size of the largest angle. (2)

10 Solve the following simultaneous equations: $2x - 3y = 7$
$$-3x + 4y = -11.$$ (3)

11  The area of this quarter circle is 50.26 cm^2.
Calculate the length of the radius. (3)

12 Villages A, B and C lie on the edge of the Sahara Desert. Village A is 29 km due north of village C, village B is 62 km due east of A.
Calculate the shortest distance between villages C and B, giving your answer to the nearest 100 metres. (3)

13 Find the total surface area of a cube of side 3 cm. (2)

Answers to practice Core exams

Practice Core exam 1

1 $y = 1, 2, 3, 5, 6, 7$ **2 a** $200 **b** $96 **3** 2 h 18 min
4 $CD = 10.8$ cm, $DE = 7.2$ cm **5** $950 \leq w < 1050$ **6** $14.20
7 a $24.8 \times 10^3 < 2.5 \times 10^5$ **b** 2.5×10^5 **c** 2.48×10^4
8 a Order 5 **b** 8 lines **9** 4.68 cm^2 **10 a i** Reflection in y-axis
ii Translation $\begin{pmatrix} 5 \\ -4 \end{pmatrix}$ **iii** Rotation 90° anti-clockwise about origin
b Coordinates are $(0, 1), (2, 1), (2, 5)$ **11 a** 106° **b** 200
12 a i $4x + 3y$ **ii** $4x + 3y = 80$ **b** $6x + 9y = 174$
c i Football shirt costs $11 **ii** Rugby shirt costs $12

Total marks = 43

Practice Core exam 2

1 a 0.75 **b** $\frac{3}{4}$ **2** 2.5 **3** $y = 5$ **4** 13 **5 a** 3.9 cm **b** 61°
6 a $60.80 **b** $72.96 **7** Fahad receives $150, Madiha receives
$200 **8 a** 17 **b** 0.25 **9** $\frac{y - c}{a} = x$ **10 a** $5x(y + 2)$
b $x(x - 4)$ **11** Cork moves from left to right, etc. forming an arc
12 a Values are 7 and 2 **b** Semi-circle expected **c i** $x = 3.5$
and 0.6 **ii** -2 **d** Values are 7 and 0 **e** Straight line graph cuts
previous graph **f** $A = (-1, 5.9)$, $B = (3.7, 1)$
13 a Frequencies: 1, 4, 4, 3, 11, 10, 7, 5, 3, 2 **b** 5.7 **c** 5
d Mode = 5, Range = 9 **e** Height of bars equals frequencies,
chart should be labelled and titled, with equal intervals for both axes.

Total marks = 54

Practice Core exam 3

1 a 38 **b** 13 **2** 15% **3** $\frac{9}{20} < 0.455 < 46\% < \sqrt{0.25}$ **4 a** RBR,
RRB **b** $\frac{2}{3}$ **5 a** 50 °C **b** 49.7 °C **6 a** 83 or 89 **b** e.g. $\sqrt{2}$ or π
c e.g. -4.5 **7 a** $\begin{pmatrix} 1 \\ -2 \end{pmatrix}$ **b** $\begin{pmatrix} 11 \\ 6 \end{pmatrix}$ **8** $a = 3, b = -2$
9 a Triangular prism **b** 84 cm^2 **c** $36\,000 \text{ mm}^3$ **10 a** £6535.95
b €3880 **11 a** 5.29 cm **b** 48.6° **12** 645 m **13 a** 90 min
b 20 km/h **c** 8 km/h **14 d** $DX = 3.7$ cm **15 a** Area of rectangle
minus two circle areas $= (12 \times 8) - 2\pi r^2$ **b i** 89.7 cm^2
ii 93.4% **c** $r = 0.98$ cm

Total marks = 56

Practice Core exam 4

1 140° **2** 18 cans **3** 720° **4 a i** 15, 18 **ii** 1.5, 0.75 **iii** 99, 182
b i 904 **ii** 11th term **c i** $\frac{6}{9}$ **ii** $\frac{17}{20}$ **iii** 20th term **iv** $\frac{n + 1}{n + 4}$
5 i 5 cm **ii** 10 cm^2 **6 a** $x - 18$ **b** $-1, 0, 1, 2, 3, 4$ **c i** 64
ii -36 **d** $x = 13$ **7 a** 90° **b** 48° **c** 90° **8 a** 24 **b** 1 **c** 2
d 2.25 **9 a** 64 cm^3 **b** 462 **10 a** $\begin{pmatrix} 6 \\ -6 \end{pmatrix}$ **b** $\begin{pmatrix} 18 \\ -18 \end{pmatrix}$
11 b $x = 2, y = 3$ **12 a** 0.9 km **b** 10 cm **13** 1 line, rotation of
order 1 **14 a** Coordinates $(1, -1), (2, -1), (2, -3)$ **b** $(-1, -1)$,
$(-2, -1), (-2, -3)$ **c** $(2.5, 1), (3, 1), (3, 2)$ **15** 2006 (after 4 years)

Total marks = 63

Practice Core exam 5

1 $\dfrac{dex + c^2}{3} = a$ **2 a** $\frac{1}{81}$ or equivalent **b** 1 **c** 1024

3 a $2x(x - 4y^2)$ **b** $7a^2b^2c(4a + a^2bc - 2c^2)$ **4** $y = \frac{1}{2}$

5 $249.5 \leq l < 250.5,\ 164.5 \leq w < 165.5$ **6 a** \$10.75 **b** 20%

7 a Values: -4, 5 and -1.25 **b** Inverted U-shape expected

c i $x = -3.2$ and 1.3 **ii** 5 **e** $x = -4$ and 1 **8 a** 1, 2, 3, 4, 6, 9, 12, 18, 36 **b** $2^2 \times 3^2$ **c** 12 **d** 72 **e** 72 000 **9** $150°$

10 $x = 5, y = 1$ **11** 8 cm **12** 68.4 km **13** 54 cm^2

Total marks = 50

Practice exams

These exams contain Paper 2 and Paper 4 type questions. The marks for each part of the question are given on the right-hand side.

Practice extended exam 1

1 Given that $4 \leq x \leq 7$ and $57 \leq y \leq 64$, find:
   a the greatest value of $\frac{y}{x}$ **b** the least value of $\frac{y}{x}$.   **(2)**

2 The value of a house increases by 20% each year. It is valued at \$20 000 today.
   a What will it be worth in two years time?   **(1)**
   b What was it worth one year ago?   **(2)**

3 Solve the quadratic equation $x^2 + 2x = 1.35$. Give your answers to 2 decimal places.   **(5)**

4 Find all the angles in the triangle.

     (4)

5 **a** Factorise $x^2 + 8x - 33$.   **(2)**

    b Simplify $\dfrac{3x + 4}{x^2 + x - 6} - \dfrac{1}{x + 3}$.   **(3)**

6 The coordinates of the points X and Y are $(-3, -1)$ and $(2, 1)$ respectively. Find:
   a the gradient of the line XY   **(2)**
   b the length of the line segment XY.   **(2)**

7 $f(x) = \dfrac{4x - 2}{5}$.
    a Find $f(4)$.   **(1)**
    b Solve $f(x) = 2$.   **(1)**
    c Find $f^{-1}(x)$.   **(2)**

8 y is inversely proportional to x^2. If $y = 1$ when $x = 0.4$, find:
   a the constant of proportionality   **(1)**
   b the value of y when $x = 0.2$ **(1)**
   c the value of x when $y = 64$.   **(1)**

9

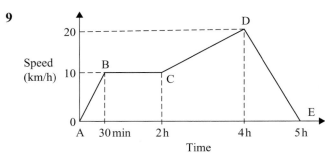

Describe this speed–time graph in *as much detail as possible*.   **(6)**

10

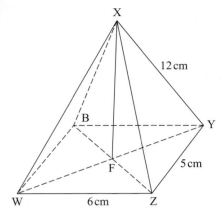

The diagram shows a rectangular-based pyramid where X is vertically above F.

  a   Calculate the length BZ.   **(2)**

  b   Calculate the length XF.   **(2)**

  c   Calculate the angle XWF.   **(2)**

11

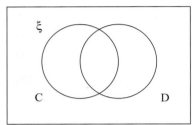

  a   Shade the region in the Venn diagram which represents the subset $C \cap D'$.   **(2)**

  b   Out of a group of 35 girls, 18 play netball, 16 play rounders and 4 play neither netball nor rounders. Find the number of girls who play rounders but not netball.   **(3)**

12   A student measures the diameter, d, of a wheel as 24 cm to the nearest centimetre.

  a   Copy and complete the statement:

$$\underline{\quad\quad} \text{ cm} \leq d < \underline{\quad\quad} \text{ cm} \qquad \textbf{(1)}$$

  b   He knows that the value of π is such that $3.0 < \pi < 3.2$. Using this information, copy and complete his statement about the circumference, C:

$$\underline{\quad\quad} \text{ cm} < C < \underline{\quad\quad} \text{ cm} \qquad \textbf{(2)}$$

13

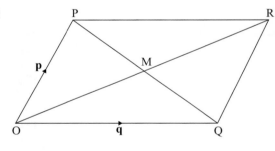

OPRQ is a parallelogram.

$\mathbf{p} = \overrightarrow{OP}$ and $\mathbf{q} = \overrightarrow{OQ}$.

Write in terms of $\mathbf{p}$ and/or $\mathbf{q}$:

  a $\overrightarrow{OR}$ **(1)**

  b $\overrightarrow{RO}$ **(1)**

  c $\overrightarrow{QP}$ **(1)**

  d $\overrightarrow{QM}$ **(1)**

14

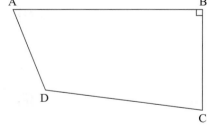

AB = 12 cm, AD = 6 cm, DC = 10 cm and BC = 8 cm. Angle B = 90°.

  a   Copy the diagram to scale.   **(2)**

  b   Draw the locus of points which are equidistant from points A and B.   **(2)**

  c   Draw the locus of points which are equidistant from lines AB and BC.   **(2)**

  d   Label the intersection of the two loci X. Measure the distance DX.   **(2)**

15   Three numbers are multiplied together.

The first number is called x, the second number is two greater than x and the third number is two less than x. If the result is zero, form an equation and find the value of x.   **(3)**

16 *Answer this question on graph paper.*

$$g(x) = \frac{12}{x}, \qquad f(x) = (x-8)(2-x)$$

x	2	3	4	5	6	7	8	9
$g(x)$	6	4	3	a	b	1.7	c	1.3
$f(x)$	0	d	8	e	8	5	0	f

    a Complete the table of values. **(3)**

    b Using a scale of 2 cm to 1 unit for the *x*-axis and 1 cm to 1 unit for the *y*-axis, draw the graphs of the above functions on the same grid. **(4)**

    c Find the gradient of the $g(x)$ curve at the point where $x = 3.5$. **(3)**

    d Find the solutions of *x* when $f(x) = 2$. **(2)**

    e Show that the equation which is satisfied by the points of intersection of the graphs can be simplified to $x^3 - 10x^2 + 16x + 12 = 0$. **(3)**

17 *Answer this question on graph paper.*

The vertices of a rectangle OPQR are (0,0), (2,0), (2,5) and (0,5) respectively.

    a Taking 1 cm to represent 1 unit on each axis and marking each axis from -6 to $+6$, draw and label the rectangle OPQR. **(3)**

    b The rectangle OPQR is mapped onto $O_1 P_1 Q_1 R_1$ by the transformation matrix $\begin{pmatrix} 0 & 1 \\ 1 & 0 \end{pmatrix}$.

       Draw and label the rectangle $O_1 P_1 Q_1 R_1$ on your diagram, describing the transformation. **(2)**

    c Reflect the original rectangle OPQR in the *y*-axis. Label it $O_2 P_2 Q_2 R_2$. **(2)**

    d Write down the matrix which represents this transformation. **(2)**

Practice extended exam 2

1 Calculate: **a** 2^{-4} **(1)**   **b** y^0 **(1)**   **c** $4^2 \div 2^{-2}$. **(1)**   **=(3)**

2 Each interior angle of a regular polygon is $157.5°$.
Calculate the number of sides of the polygon. **(2)**

3 Solve for *x* and *y*: $5x + 2y = 13$
                     $2x - y = 7$. **(4)**

4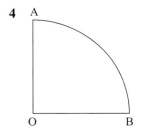

    The area of this quarter circle is 28.28 cm^2.
    Calculate the length of the radius. **(3)**

5 **a** Factorise $3x^2 - 5x - 2$. **(2)**

    b Simplify $\dfrac{2}{x} - \dfrac{1}{x+3}$. **(2)**

6 A bank charges $42 simple interest when $700 is borrowed for 6 months.
Calculate the annual percentage interest rate. **(3)**

7 $x = \dfrac{2y - z^2}{ab}$. Make *z* the subject of the formula. **(3)**

8 $A = \begin{pmatrix} 4 & 3 \\ 2 & 1 \end{pmatrix}$ $B = \begin{pmatrix} -2 & -4 \\ 3 & 6 \end{pmatrix}$ $C = \begin{pmatrix} 2 & 4 & 6 \\ 1 & 3 & -3 \end{pmatrix}$

    Calculate: **a** A^2 **(1)**   **b** AB **(1)**   **c** BC **(1)**   **d** $A^{-1} B^{-1} C$. **(2)**   **=(5)**

9 A map has a scale of 1 : 40 000.
  a A road on the map is 10 cm long. What is the real length of the road in kilometres? **(1)**
  b The area of a farm on the map is $9\,\text{cm}^2$. What is the real area of the farm in hectares?
    (One hectare $= 10\,000\,\text{m}^2$.) **(2)**

10

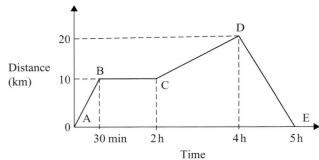

Describe this distance–time graph in *as much detail as possible*. **(5)**

11 A water tank is in the shape of a cuboid, measuring 150 cm by 100 cm by 30 cm.
  a How many litres of water would the tank contain when full? **(2)**
  b The tank is initially empty and water flows into it from a pipe.
    The cross-sectional area of the pipe is $2.4\,\text{cm}^2$ and the water flows along the pipe at a rate of 35 cm/s. By calculating the volume of water flowing from the pipe in 1 second, find the time taken (to the nearest minute) to fill the tank. **(4)**

12 The table below shows the amount of money, x, spent on books by a group of students.

Amount spent, x	$0 < x \le 10$	$10 < x \le 20$	$20 < x \le 30$	$30 < x \le 40$	$40 < x \le 50$	$50 < x \le 60$
Number of students	0	4	8	12	11	5

  a Calculate an estimate of the mean amount of money spent per student on books. **(4)**
  b Construct a cumulative frequency curve from the information in the table. **(5)**
  c Use the curve to find:
    i the median amount spent **(2)**
    ii the interquartile range. **(2)**

13 *Answer this question on graph paper*.
    The vertices of a triangle ABC are (1, 2), (4, 2) and (4, 4) respectively.
  a Taking 1 cm to represent 1 unit on each axis and marking each axis from -6 to $+ 6$, draw and label the triangle ABC. **(3)**
  b The triangle ABC is rotated $90°$ anti-clockwise about the origin onto $A_1 B_1 C_1$. Draw and label $A_1 B_1 C_1$. **(2)**
  c Find the matrix which represents this transformation. **(2)**
  d Triangle $A_1 B_1 C_1$ is reflected in the x-axis onto $A_2 B_2 C_2$. Draw and label $A_2 B_2 C_2$. **(2)**
  e Find the matrix which represents this transformation. **(2)**

Practice extended exam 3

1 Calculate: **a** $8 + 7 \times 3$ **(1)** **b** $\frac{1}{3}(9 \times 4 - 6 \div 2)$. **(2)** =**(3)**

2 A regular hexagon has sides of 4.5 cm. Calculate its area. **(4)**

3 Solve for c and d: $3c + 4d = -17$ $2c - 2d = -2$. **(4)**

4

A and B are both spheres. Sphere A has a radius of 8 cm and sphere B has a radius of 2 cm. Find by calculation:
  a the volume of sphere A **(2)**
  b the surface area of sphere B. **(2)**
  c Hence prove that the shapes are similar. **(2)**

Surface area of sphere $= 4\pi r^2$

Volume of sphere $= \frac{4}{3}\pi r^3$

5 Factorise completely: **a** $16x^2 - 121y^2$ **(2)**
                                           b $ax + ay + bx + by$.   **(2)**

6 A lorry has a capacity of 50 000 litres. A model of the lorry is made using a scale of 1 : 40.
Calculate the capacity of the model in millilitres.   **(3)**

7 $a = \left(\dfrac{4y + b^2}{gh}\right)^2$ Make b the subject of the formula.   **(3)**

8 The vector $\mathbf{a} = \begin{pmatrix} 7 \\ 2 \end{pmatrix}$ and $\mathbf{b} = \begin{pmatrix} -3 \\ -1 \end{pmatrix}$. Find:

    a $\mathbf{a} + \mathbf{b}$ **(1)**   **b** $2\mathbf{a} - \mathbf{b}$ **(2)**   **c** $|\mathbf{a}|$, the magnitude of $\mathbf{a}$.   **(2)**   **=(5)**

9 A map has a scale of 1 : 250.
    a   A house on the map is 4 cm long, what is the real length of the house in metres?   **(1)**
    b   The area of a lake on the map is 10 cm^2. What is the real area of the lake in metres^2 ?   **(2)**

10

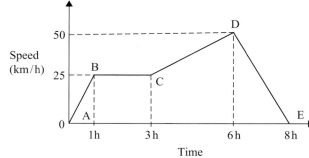

Describe this speed–time graph
in *as much detail as possible*.   **(6)**

11 A train usually completes a journey of 180 km at an average speed of x km/h. One day
engine trouble causes the average speed to be 20 km/h less than usual.
    a   Write down an expression for the time taken, h, for the usual journey.   **(2)**
    b   Write down an expression for the time taken, h, for the slower journey.   **(2)**
    c   The time for the slower journey was 15 minutes more than the time for the usual journey.
       Write down an equation for x using your answers to parts **a** and **b**, and hence show that:

$$x^2 - 20x - 14\,400 = 0$$ **(4)**

    d   Solve the equation $x^2 - 20x - 14\,400 = 0$ and hence calculate the usual average speed
       of the train.   **(4)**

12 *Answer this question on graph paper.*
The table below shows the height of flowers as measured during an experiment.

Height (cm)	$0 < h \le 5$	$5 < h \le 10$	$10 < h \le 15$	$15 < h \le 25$	$25 < h \le 50$
Frequency	20	40	60	80	50

    a   Calculate an estimate of the mean height of the flowers.   **(4)**
    b   Construct a histogram from the information in the table. (Choose your own scale.)   **(5)**
    c   If two flowers are picked at random, state the probability that both flowers have a height
       greater than 10 cm.   **(2)**

13 *Answer this question on graph paper.*
The vertices of a rectangle ABCD are (1, 2), (4, 2), (4, 4) and (1,4) respectively.
    a   Taking 1 cm to represent 1 unit on each axis and marking each axis from -10 to $+10$,
       draw and label the rectangle ABCD.   **(3)**
    b   The rectangle ABCD is rotated 90° clockwise about the origin onto $A_1\,B_1\,C_1\,D_1$.
       Draw and label $A_1\,B_1\,C_1\,D_1$.   **(2)**
    c   Find the matrix which represented this transformation.   **(2)**

d Rectangle $A_1 B_1 C_1 D_1$ is reflected in the *y*-axis onto $A_2 B_2 C_2 D_2$.
Draw and label $A_2 B_2 C_2 D_2$. **(2)**

e Rectangle $A_2 B_2 C_2 D_2$ is enlarged by scale factor 2, using the origin as centre of
enlargement, onto $A_3 B_3 C_3 D_3$. Draw and label $A_3 B_3 C_3 D_3$. **(2)**

Practice extended exam 4

1 a If 25% of the 350 million sheep in the world were bred in Wales, calculate the number of
sheep in Wales. **(1)**

b If 45% of Scotsmen wear a skirt (kilt), what is the probability that Mr McDuff or
Mr McHaggis wears a skirt at home? **(2)**

c Mr Voice can eat 4 pizzas every day, yet still manages to maintain his athletic physique.
If each pizza contains 500 calories, calculate how many calories he consumes in one
year. (Assume it's not a leap year.) **(2)**

2 An irregular heptagon has 4 equal angles of 105°. If the remaining three angles are in the
ratio 1 : 2 : 3, calculate the largest angle. **(3)**

3 Arrange the following in order of size, smallest first: 0.5 m², 450 cm², 45 500 mm². **(2)**

4 Describe *fully* the symmetry of this shape: **(2)**

5 Given the functions $f(x) = x^2 + 6$ and $g(x) = 2x + 1$, calculate:
a $f(2)$ **(1)**   **b** $g(-3)$ **(1)**   **c** $g^{-1}(x)$ **(2)**   **d** $fg^{-1}(2)$ **(2)**   **=(6)**

6 Given that Mr McTurk can save 95% of his salary every month and that his annual savings
amount to $23 750, what is his yearly salary? **(2)**

7

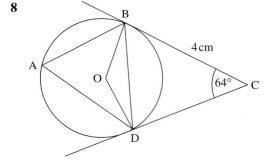

Mr De Jong

Mr Gray

1.5 m

1.2 m

Mr Gray and Mr De Jong are similar in
shape.
a If Mr Gray weighs 100 kg, what is the
weight of Mr De Jong? **(2)**
b If they were both rolled flat by a passing
steamroller and Mr Gray covered a
surface area of 10 m², what surface area
would Mr De Jong cover? **(2)**

8

B

4 cm

A

O

64° C

D

BC and CD are tangents to the circle,
centre O.
Angle BCD = 64°, BC = 4 cm.
Find:
a angle BAD **(1)**
b angle CBD **(1)**
c angle BOD **(1)**
d chord length BD. **(3)**

9 a Draw the *accurate* net of the solid shown. **(3)**
b Calculate its volume and surface area. **(2), (2)**

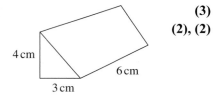

4 cm

6 cm

3 cm

10 Fahad desperately tries to get a question right in Maths. Each time he tries, the probability that he succeeds is 0.2. Each time he fails, he tries again.

   a Copy and complete the tree diagram below: **(2)**

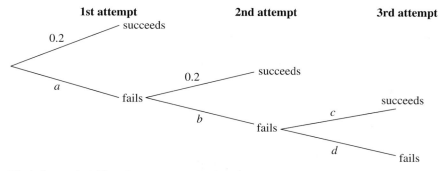

   b Find the probability that, to succeed, it takes:
     i exactly two tries **(2)**
     ii one, two or three tries **(3)**
     iii n tries. **(2)**

11 *Answer this question on graph paper.*

$$f(x) = 4 - x - x^2 \qquad g(x) = x^3 + 1$$

The tables below show some values for both functions:

x	-4	-3	-2	-1	0	1	2	3
$f(x)$	a	b	c	4	4	2	d	e

x	-2	-1.5	-1	-0.5	0	0.5	1	1.5
$g(x)$	-7	f	g	h	1	i	2	j

   a Complete the tables for both functions. **(3)**
   b Draw both graphs on the same grid. **(4)**
   c Use your graph(s) to solve the equation $4 - x - x^2 = 0$. **(2)**
   d Estimate the gradient of the $f(x)$ function where $x = -2$. **(3)**
   e At which x value(s) does $f(x) = g(x)$? **(2)**

Practice extended exam 5

1
Mr Mackenzie uses a large 'Jumbo pack' of black hair dye every week.
Calculate:
   a the cost per litre **(2)**
   b the number of millilitres of hair dye obtained for one dollar **(2)**
   c the cost price of the tin if the shopkeeper made a 20% profit on its sale. **(2)**

2 By splitting into triangles (no overlapping), prove that an irregular dodecagon obeys the formula for calculating the total degrees inside any polygon. **(4)**

3 Factorise: **a** $6x^2 - 9x - 27$ **(2)**   **b** $144a^2 - 16b^2$ **(2)**   **c** $3 - 75y^2$. **(2)**   **=(6)**

4 **a** In a class, two students are chosen at random from 14 boys and 11 girls. Find the probability that:
     i two boys are chosen **(2)**   **ii** a boy or a girl is chosen. **(2)**   **=(4)**
   b Express 340 as the product of its prime factors. **(2)**

5 a Madiha and Jessica share a sum of money in the ratio 5 : 3. If Madiha's share is $21.25, find the total amount of money they have to share. **(2)**

   b Mr Horsey enjoys blowing his own trumpet. If the mass of his trumpet is 4.8 kg and represents 4% of his weight, find the combined mass of both Mr Horsey and his trumpet. **(2)**

6 Simplify $(a^{-1} b^0 c^2 d^{-3}) \times (2abc^{-2}d)$ giving your answer in its lowest terms. **(2)**

7

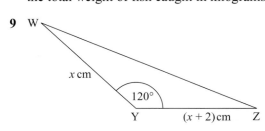

In a recent teacher popularity survey involving 720 pupils, the results were recorded and displayed as a pie chart.

   a What fraction of voters voted for Mr Morse ? (Lowest terms) **(1)**

   b How many pupils voted for Mr Law? **(1)**

   c What percentage of pupils voted for Mrs Meyer? **(1)**

8 On a recent teachers' fishing trip, Mr Law caught two large sharks, Mr Gray caught two tiny sardines and Mr Benfield foul-hooked an elderly catfish half the size of a shark. Given that the weight ratio of shark : sardine was 1500 : 1 and the weight of a sardine was 3 g, calculate the total weight of fish caught in kilograms. **(3)**

9

   a Use the cosine rule to find an expression for WZ^2 in terms of x. **(4)**

   b When WZ is 7 cm, show that $2x^2 + 4x - 30 = 0$. **(3)**

   c Factorise $2x^2 + 4x - 30$. **(2)**

   d Solve the equation $2x^2 + 4x - 30 = 0$ and hence write down the lengths of WY and YZ. **(4)**

10 Jim and Fred are tailors. They make x jackets and y suits each week. Jim does all the cutting and Fred does all the sewing. To make a jacket takes 5 hours of cutting and 4 hours of sewing. To make a suit takes 6 hours of cutting and 10 hours of sewing. Neither tailor works for more than 60 hours a week.

   a For the sewing, show that $2x + 5y \le 30$. **(2)**

   b Write down another inequality in x and y for the cutting. **(1)**

   c They make at least 8 jackets each week. Write down another inequality. **(1)**

   d Draw axes from 0 to 16, using 1 cm to 1 unit on each axis, and on your grid show the information in parts **a**, **b** and **c**, shading the unwanted regions. **(6)**

   e The profit on a jacket is $40 and on a suit $110. Calculate the maximum profit they can make in a week. **(3)**

Practice extended exam 6

1. Give an example of:

   a a negative number which is not an integer **(1)**

   b a prime number between 60 and 70 **(1)**

   c any irrational number. **(1)**

2

   a Shade $A' \cup B'$ on a copy of the diagram. **(2)**

   b In a class of 44 students, 20 study Physics, 28 study Chemistry and 3 study neither. How many study both Physics and Chemistry? **(3)**

3 To the nearest metre, a room measures 8 metres long by 6 metres wide.
   a The actual length of the room is l metres. Write down the upper and lower limits of l. **(1)**
   b The actual area of the room is A metres^2. Calculate the upper and lower limits of A. **(2)**

4

   a Calculate AB, the distance from town A to town B. **(2)**
   b Calculate angle BAC to the nearest degree. **(2)**
   c Find the bearing of town A from town B. **(2)**
   d A plane flew from town A to town B at an average speed of 400 km/h. If it left town A at 0840 hours, at what time did it arrive in town B? **(2)**

5 Factorise: **a** $2x^2 - 8xy^2$ **(2)**   **b** $2y^2 + 7y + 6$. **(2)**   =**(4)**

6 Solve the equation $3y + 17 = 21 - 5y$. **(2)**

7 Make x the subject of the formula $y = \dfrac{c + 3x}{2}$. **(3)**

8 The values a and b satisfy the equation $a^2b = k$, where k is a constant number.
   a If $a = 3$ when $b = 10$, find the value of b when $a = 2$. **(2)**
   b What happens to the value of b when the value of a is decreased by 50%? **(2)**

9 *Answer the whole of this question on graph paper.*
Ahmed, Ali and Hanni sell houses.
Ahmed charges \$500, whatever the selling price. Ali charges 1% of the selling price. Hanni charges \$250 for selling prices up to \$30 000. For selling prices more than \$30 000, he charges \$250 plus 2% of the value over \$30 000.
   a Use a scale of 2 cm to represent a selling price of \$10 000 on the horizontal axis and 2 cm to represent a charge of \$100 on the vertical axis.
On the same grid, draw the three graphs to show the charges made by Ahmed, Ali and Hanni for selling prices up to \$80 000.
Label your graphs clearly. **(5)**
   b **i** For which selling price is Ahmed's charge the same as Ali's ? **(2)**
      ii For what range of selling prices does Hanni charge the least? **(2)**
     iii For which selling price less than \$50 000 does Ali charge \$50 less than Hanni? **(2)**

10

In the diagram, O is the origin, YZW is a straight line and X is the mid-point of OY.

$$\overrightarrow{OY} = \mathbf{a} \qquad \overrightarrow{OZ} = \mathbf{b} \qquad \overrightarrow{YW} = 3\overrightarrow{YZ}$$

Find in terms of $\mathbf{a}$ and/or $\mathbf{b}$ in their simplest form:
   a $\overrightarrow{XY}$ **(2)**   **b** $\overrightarrow{YZ}$ **(2)**   **c** $\overrightarrow{XW}$ **(2)**   **d** the position vector of W. **(2)**   =**(8)**

   e Given also that $\overrightarrow{XV} = \frac{1}{5}\overrightarrow{XW}$, find:
    i $\overrightarrow{OV}$ in terms of $\mathbf{a}$ and/or $\mathbf{b}$ **(2)**
    ii the ratio $\overrightarrow{OV} : \overrightarrow{VZ}$. **(2)**

Practice extended exam 7

1 Solve the equation $5 - \dfrac{x}{0.5} = -11$. **(2)**

2 Solve for a and b $4a - b = 11$
$3a - 3b = 15.$ **(3)**

3 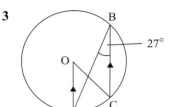 O is the centre of the circle and lines BC and OA are parallel.
Given that angle ABC is $27°$, calculate:

  a angle OAB **(2)**

  b angle AOC. **(2)**

4 **a** Factorise $4 - 100a^2$. **(2)**

  b Simplify $\dfrac{2}{x^2 + 5x + 6} - \dfrac{1}{x + 3}$ **(3)**

5 Natasha and Sarosh share a sum of money in the ratio 5 : 4. If Natasha's share is $22.50, find the total amount of money shared altogether. **(2)**

6

A ————— 10 cm ————— D
5 cm
B ————— 8 cm ————— C

  a Draw this shape to scale. **(2)**

  b Construct the locus of points which are 6 cm from D. **(2)**

  c Construct the locus of points which are equidistant from A and B. **(2)**

  d Shade the region which is less than 6 cm from D and nearer to B than A. **(2)**

7 A high-speed train usually completes a journey of 400 km at an average speed of x km/h. However, on one snowy day, ice on the track causes its average speed to be 20 km/h less than usual.

  a Write down an expression, in terms of x, for the time taken in hours. **(1)**

  b Write down a similar expression for the time taken for its slower journey. **(1)**

  c If the difference in time between the journeys was 45 minutes, write an equation in x and hence show that $3x^2 - 60x - 32\,000 = 0$. **(4)**

  d Solve the equation $3x^2 - 60x - 32\,000 = 0$, and hence write down the usual average speed of the train. **(5)**

8 Ten counters are placed in a bag, of which three are red, five are blue and two are green. Random selections are made **without replacement**.

  a Copy and complete the probability tree below for values a to d. **(2)**

  b Calculate the probability of obtaining:

    i a red and then a blue **(2)**

    ii two of the same colour **(3)**

    iii three blues in a row. **(2)**

2nd selection

1st selection

$P\,(Red) = \dfrac{2}{9}$

$P\,(Blue) = c$

$P\,(Red) = \dfrac{3}{10}$

$P\,(Green) = d$

$P\,(Blue) = a$

$P\,(Green) = b$

9

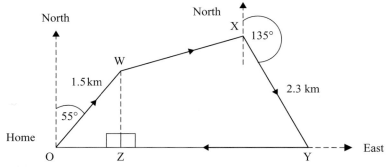

The diagram shows the path taken by a walker one sunny afternoon.
Starting from home, he walks to W then X, then Y and back to his house.
All distances are in kilometres.

a Calculate the distances $\overrightarrow{OZ}$ and $\overrightarrow{ZW}$, giving answers to one decimal place. **(2), (2)**

b Write down the vector $\overrightarrow{OW}$ in the form $\begin{pmatrix} x \\ y \end{pmatrix}$, where x and y are the components of the

  column vector. **(1)**

c $\overrightarrow{WX} = \begin{pmatrix} 2 \\ 0.75 \end{pmatrix}$ Calculate: **i** length of WX, to 1 decimal place **(2)**
  ii bearing of X from W, to the nearest degree. **(2)**

d Calculate the total distance walked. **(3)**

Practice extended exam 8

1 Prove with the use of two examples that Pythagoras' theorem will only work on a
right-angled triangle. **(3)**

2 The pressure, P, of a gas is inversely proportional to its volume, V, at a constant temperature.
If $P = 4$ when $V = 6$, calculate:
a P when $V = 8$ **(2)**
b V when $P = 12$. **(2)**

3

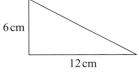

The triangle is to be reduced by a ratio of $1:3$.
a Calculate the area of the original triangle. **(1)**
b Calculate the area of the reduced triangle. **(2)**
c Calculate the ratio by which the area of the triangle has
been reduced. **(2)**

4

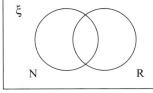

a Shade the region in the Venn diagram which represents
the subset $N' \cap R$. **(2)**
b Out of a group of 25 girls, 15 play netball, 10 play
rounders and 4 play neither netball nor rounders.
Find the numbers of girls who play rounders but not
netball. **(3)**

5 $\xi = \{x : 3 \leq x \leq 12 \text{ and } x \text{ is an integer}\}$, $C = \{\text{prime numbers}\}$, $D = \{\text{factors of } 12\}$.
a List the sets: **i** $C \cap D$ **(2)** **ii** $C \cup D$ **(2)** $=$**(4)**
b State the value of $n(D')$. **(2)**

6 Calculate: a length of side BC. **(3)**
  b area of triangle ABC. **(2)**

7 If $f(x) = 4x - 6$, calculate:

    a $f(3)$ **(1)**   **b** $f(x) = 6$ **(2)**   **c** $f^{-1}(x)$ **(2)**   **d** $f^{-1}(-2)$ **(1)**   =**(6)**

8 The spherical ball below is floating in a liquid with its centre O at a depth h below the surface. XZ is a diameter of the circular cross-section formed at the surface. When $r = 26$ cm and $h = 10$ cm,

    a Calculate:

        i Length of XZ **(2)**   **ii** Angle XOZ **(2)**   **iii** Length of the arc XYZ. **(2)**   =**(6)**

    b The area of surface above the liquid level is given by the formula $2\pi r (r - h)$.

        Find the area of surface above the liquid level.   **(2)**

    c If the total surface area of a sphere is given by the formula $4\pi r^2$, find the area above the liquid level as a percentage of the total surface area of the sphere.   **(3)**

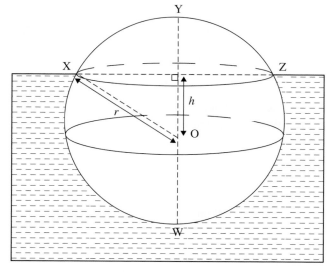

9 *Answer this question on graph paper.*

The vertices of a triangle ABC are (1, 1), (3, 1), (3, 4) respectively.

    a Taking 1 cm to represent 1 unit on each axis and marking each axis from −8 to +10, draw and label the triangle ABC.   **(3)**

    b The triangle ABC is reflected in the y-axis onto $A_1 B_1 C_1$.

        i Draw and label $A_1 B_1 C_1$.   **(2)**

        ii Find the matrix which represents this transformation.   **(2)**

    c Triangle $A_1 B_1 C_1$ is enlarged from point X (1, −1) by a scale factor of 2 onto triangle $A_2 B_2 C_2$. Draw and label $A_2 B_2 C_2$.   **(2)**

    d Triangle $A_2 B_2 C_2$ is then translated by the vector $\begin{pmatrix} 0 \\ -7 \end{pmatrix}$ to its new position $A_3 B_3 C_3$.

        Draw and label $A_3 B_3 C_3$.   **(2)**

    e A shear transformation of $A_1 B_1 C_1$, shear factor 2, direction parallel to the x-axis, with $A_1 B_1$ as the invariant line, is then performed. Draw and label this transformation $A_4 B_4 C_4$.   **(2)**

Answers to practice Extended exams

Practice Extended exam 1

1 a 16   **b** 8.14   **2 a** \$28 800   **b** \$16 666.67   **3** $x = 0.53$ or
-2.53 **4** $\widehat{W} = 87.3°$, $\widehat{X} = 51°$, $\widehat{Y} = 41.7°$ **5 a** $(x + 11)(x - 3)$
b $\dfrac{2}{x - 2}$ **6 a** $\frac{2}{5}$ **b** 5.4   **7 a** 2.8   **b** $x = 3$ **c** $\dfrac{5x + 2}{4}$
8 a $k = 0.16$ **b** $y = 4$ **c** $x = 0.05$ **9** A to B = constant
acceleration of 20 km/h^2, B to C = constant speed of 10 km/h, C to
D = constant acceleration of 5 km/h^2, D to E = constant
deceleration of 20 km/h^2, total distance travelled = 57.5 km
10 a 7.8 cm   **b** 11.3 cm   **c** 71°   **11 a** Shade all of C except the
intersection   **b** 13 girls   **12 a** $23.5 \le d < 24.5$
b $70.5 < C < 78.4$ **13 a** $\mathbf{p + q}$ **b** $\mathbf{-p - q}$ **c** $\mathbf{p - q}$
d $\frac{\mathbf{p-q}}{2}$ **14** All construction arcs should be shown **d** DX = 4.8 cm
15 $x(x + 2)(x - 2) = 0$, $x = 0$, 2 or -2 **16 a** $a = 2.4$, $b = 2$,
$c = 1.5$, $d = 5$, $e = 9$, $f = -7$ **c** -1 **d** $x = 2.3$ or 7.8
e Rearrange algebraically $f(x) = g(x)$ **17 b** Coordinates of
rectangle are O$_1$ (0, 0), P$_1$ (0, 2), Q$_1$ (5, 2), R$_1$ (5, 0), reflection in
$y = x$ **c** O$_2$ (0, 0), P$_2$ $(-2, 0)$, Q$_2$ $(-2, 5)$, R$_2$ (0, 5)   **d** $\begin{pmatrix} -1 & 0 \\ 0 & 1 \end{pmatrix}$

Total marks = 89

Practice Extended exam 2

1 a $\frac{1}{16}$ **b** 1   **c** 64   **2** 16 sides   **3** $x = 3$, $y = -1$ **4** 6 cm
5 a $(3x + 1)(x - 2)$ **b** $\dfrac{x + 6}{x(x + 3)}$ **6** 12%   **7** $z = \sqrt{2y - abx}$

8 a $\begin{pmatrix} 22 & 15 \\ 10 & 7 \end{pmatrix}$ **b** $\begin{pmatrix} 1 & 2 \\ -1 & -2 \end{pmatrix}$ **c** $\begin{pmatrix} -8 & -20 & 0 \\ 12 & 30 & 0 \end{pmatrix}$

d Impossible, determinant = 0 so no inverse   **9 a** 4 km
b 144 hectares   **10** AB = constant speed of 20 km/h (1 mark),
BC = stationary for 1.5 hours (1 mark), CD = constant speed of
5 km/h (1 mark), DE = returning to start point at constant speed of
20 km/h (2 marks)   **11 a** 450 litres   **b** Volume = 84 cm^3, time =
5357 seconds = 89 minutes (to nearest minute)   **12 a** \$36.25
b Ensure cumulative frequency plotted against upper class limit
c i Median = \$38   **ii** Interquartile range = $45 - 28 = \$17$
13 b Coordinates of triangle are A$_1$ $(-2, 1)$, B$_1$ $(-2, 4)$, C$_1$ $(-4, 4)$.
c $\begin{pmatrix} 0 & -1 \\ 1 & 0 \end{pmatrix}$ **d** A$_2$ $(-2, -1)$, B$_2$ $(-2, -4)$, C$_2$ $(-4, -4)$

e $\begin{pmatrix} 1 & 0 \\ 0 & -1 \end{pmatrix}$

Total marks = 65

Practice Extended exam 3

1 a 29   **b** 11   **2** $6(\frac{1}{2} \times 4.5^2 \times \sin 60°) = 52.6$ cm^2 **3** $c = -3$,
$d = -2$ **4 a** 2145 cm^3 **b** 50.3 cm^2 **c** Use ratios of $(8 : 2)^2$ and
$(8 : 2)^3$ to show connection between surface areas and volumes
5 a $(4x - 11y)(4x + 11y)$ **b** $(a + b)(x + y)$
6 $\dfrac{50\,000}{40^3} = 781$ ml   **7** $b = \sqrt{(\sqrt{a}\,gh - 4y)}$ **8 a** $\begin{pmatrix} 4 \\ 1 \end{pmatrix}$ **b** $\begin{pmatrix} 17 \\ 5 \end{pmatrix}$

c 7.3 units   **9 a** 10 m   **b** 62.5 m^2 **10** AB = constant acceleration
of 25 km/h^2 (1 mark), BC = constant speed of 25 km/h (1 mark),
CD = constant acceleration of 8.33 km/h^2 (1 mark), DE = constant
deceleration of 25 km/h^2 (1 mark), total distance travelled = 225 km
(2 marks)   **11 a** $\dfrac{180}{x}$ **b** $\dfrac{180}{x - 20}$ **c** $\dfrac{180}{x - 20} - \dfrac{180}{x} = 0.25$ then
simplify to $x^2 - 20x - 14\,400 = 0$ **d** 130.4 km/h

12 a 18.3 cm **b** $0 < h \le 5 = 4\,\text{cm}^2$, $5 < h \le 10 = 8\,\text{cm}^2$, $10 < h \le 15 = 12\,\text{cm}^2$, $15 < h \le 25 = 16\,\text{cm}^2$, $25 < h \le 50 = 8\,\text{cm}^2$ **c** $\dfrac{190}{250} \times \dfrac{189}{249} = \dfrac{1197}{2075}$

13 b Coordinates of rectangle are A_1 $(2, -1)$, B_1 $(2, -4)$, C_1 $(4, -4)$, D_1 $(4, -1)$ **c** $\begin{pmatrix} 0 & 1 \\ -1 & 0 \end{pmatrix}$ **d** A_2 $(-2, -1)$, B_2 $(-2, -4)$, C_2 $(-4, -4)$, D_2 $(-4, -1)$ **e** A_3 $(-4, -2)$, B_3 $(-4, -8)$, C_3 $(-8, -8)$, D_3 $(-8, -2)$

Total marks = 75

Practice Extended exam 4

1 a 87.5 million **b** $\frac{9}{10}$ **c** 730 000 calories **2** $240°$ (1 mark for calculating total degrees of 900, extra mark for finding one share of total degrees) **3** $450\,\text{cm}^2 < 45\,500\,\text{mm}^2 < 0.5\,\text{m}^2$ **4** 1 line of symmetry, order 1 rotation symmetry (1 mark each) **5 a** 10 **b** -5 **c** $\dfrac{x-1}{2}$ **d** 6.25 **6** $25\,000 **7 a** 195.3 kg **b** $15.6\,\text{m}^2$ (allow a mark for stating the cube/squared ratio) **8 a** $58°$ **b** $58°$ **c** $116°$ **d** 4.24 cm **9 b** Volume $= 36\,\text{cm}^3$, surface area $= 84\,\text{cm}^2$ **10 a** $a = 0.8$, $b = 0.8$, $c = 0.2$, $d = 0.8$ ($\frac{1}{2}$ mark each) **b i** 0.16 **ii** 0.488 **iii** $(0.2)0.8^{n-1}$ **11 a** $a = -8$, $b = -2$, $c = 2$, $d = -2$, $e = -8$, $f = -2.38$, $g = 0$, $h = 0.88$, $i = 1.13$, $j = 4.38$ (1 mistake, lose 1 mark, otherwise no marks) **c** $x = -2.6$ or 1.6 **d** Gradient $= 2.8 \pm 0.5$ (allow 1 mark for tangent) **e** $x = 1$

Total marks = 60

Practice Extended exam 5

1 a $2.90 **b** 345 ml **c** $12.08 **2** 10 triangles should be present, which equals $1800°$, the same as $(n-2) \times 180°$ formula gives **3 a** $(3x - 9)(2x + 3)$ **b** $(12a - 4b)(12a + 4b)$ **c** $3(1 - 5y)(1 + 5y)$ **4 a i** $\frac{91}{300}$ **ii** 1 **b** $2^2 \times 5 \times 17$ **5** $34 (1 mark for finding Jessica's share was $12.75) **b** 124.8 kg (1 mark for finding Mr Horsey's weight was 120 kg) **6** $2bd^{-2}$ or equivalent **7 a** $\frac{5}{18}$ **b** 280 pupils **c** 33.3.% **8** 11.256 kg **9 a** $WZ^2 = 3x^2 + 6x + 4$ **b** $49 = 3x^2 + 6x + 4$, $3x^2 + 6x - 45 = 0 \,(\div 1.5)$ to give $2x^2 + 4x - 30 = 0$ **c** $(2x - 6)(x + 5)$ **d** $WY = 3\,\text{cm}$, $YZ = 5\,\text{cm}$ **10 a** $4x + 10y \le 60 \,(\div 2)$ to give $2x + 5y \le 30$ **b** $5x + 6y \le 60$ **c** $x \ge 8$ **d** Plot graphs, region left unshaded will have boundaries $(8, 0)$, $(8, 2.8)$, $(9.2, 2.3)$ and $(12, 0)$ **c** 9 jackets and 2 suits $= $580 profit

Total marks = 60

Practice Extended exam 6

1 a Any negative decimal or fraction, e.g. -2.5 **b** 61 or 67 **c** e.g. π, $\sqrt{2}$ **2 a** All regions shaded except $A \cap B$ **b** 7 **3 a** $7.5 \le l < 8.5$ **b** $41.25 \le A < 55.25$ **4 a** 1118.8 km **b** $40.6°$ **c** $218°$ **d** 1127 hours **5 a** $2x(x - 4y^2)$ **b** $(2y + 3)(y + 2)$ **6** $y = 0.5$ **7** $x = \frac{2y-c}{3}$ **8 a** $k = 90$, $b = 22.5$ **b** a quadruples or increases by 300%. **9 b i** $50 000 **ii** $25 000 to $30 000 **iii** $40 000 and $20 000 **10 a** $\frac{1}{2}a$ **b** $b - a$ **c** $3b - 2.5a$ **d** $3b - 2a$ **e i** $\frac{3}{5}b$ **ii** $3 : 2$

Total marks = 55

Practice Extended exam 7

1 $x = 8$ **2** $a = 2$, $b = -3$ **3 a** $27°$ **b** $54°$ **4 a** $(2 - 10a)(2 + 10a)$ **b** $\dfrac{-x}{(x + 2)(x + 3)}$ **5** $40.50

6 Section shaded should be below the perpendicular bisector of AB but also within the circle. This will be a small area in the bottom right of the quadrilateral. **7 a** $\dfrac{400}{x}$ **b** $\dfrac{400}{x-20}$

c $\dfrac{400}{x-20} - \dfrac{400}{x} = 0.75$ then simplify to $3x^2 - 60x - 32\,000 = 0$

d Usual speed $= 113.8$ km/h **8 a** $a = \frac{5}{10}$, $b = \frac{2}{10}$, $c = \frac{5}{9}$, $d = \frac{2}{9}$

b i $\frac{1}{6}$ **ii** $\frac{14}{45}$ **iii** $\frac{1}{12}$ **9 a** OZ $= 1.2$ km, WZ $= 0.9$ km

b $\begin{pmatrix} 1.2 \\ 0.9 \end{pmatrix}$ **c i** 2.1 km **ii** $90 - 20.6 = 069°$

d $1.5 + 2.1 + 2.3 + 1.65 + 2 + 1.23 = 10.78$ km

Total marks = 56

Practice Extended exam 8

1 E.g. choose a right-angled triangle showing that $c^2 = a^2 + b^2$, then pick a non-right angled triangle and show that it does not obey Pythagoras' theorem. **2 a** $k = 24$, $P = 3$ **b** $V = 2$ **3 a** 36 cm^2

b 4 cm^2 **c** $1 : 9$ **4 a** All of R shaded except the intersection

b 6 players **5 a i** $\{3\}$ **ii** $\{3, 4, 5, 6, 7, 11, 12\}$ **b** 6

6 a 12.2 cm **b** 44.2 cm^2 **7 a** 6 **b** 3 **c** $\dfrac{x+6}{4}$ **d** 1

8 a i 48 cm **ii** 134.8° **iii** 61.2 cm **b** 2614 cm^2 **c** 30.8%

9 b i Coordinates are $A_1 (-1, 1)$, $B_1 (-3, 1)$, $C_1 (-3, 4)$

ii $\begin{pmatrix} -1 & 0 \\ 0 & 1 \end{pmatrix}$ **c** $A_2 (-3, 3)$, $B_2 (-7, 3)$, $C_2 (-7, 9)$

d $A_3 (-3, -4)$, $B_3 (-7, -4)$, $C_3 (-7, 2)$ **e** $A_4 (-1, -1)$, $B_4 (-3, 1)$, $C_4 (3, 4)$

Total marks = 58

Formulae

Core section

Rectangle area = length × width   perimeter = 2 × length + 2 × width = (distance around shape)
Triangle   area = $\frac{1}{2}$base × height   perimeter = add up all the sides

Parallelogram   area = base × height

Trapezium   area = $\frac{\text{sum of parallel sides}}{2}$ × height

Circle   area = πr^2 circumference = $2\pi r$ where r is the radius

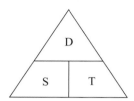

Distance = speed × time
Speed = distance ÷ time
Time = distance ÷ speed

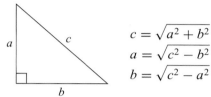

$c = \sqrt{a^2 + b^2}$
$a = \sqrt{c^2 - b^2}$
$b = \sqrt{c^2 - a^2}$

Gradient of a graph = $\frac{\text{change in } y}{\text{change in } x}$

Mean = total of values ÷ sum of values

Mode = most frequent number

Median = middle value when numbers in order of size

% increase/decrease = $\frac{\text{actual change} \times 100}{\text{original}}$

Total degrees in polygon = $(n - 2) \times 180°$

Interior angle = Total degrees ÷ n where n = number of sides

Sin $\theta = \frac{\text{opposite}}{\text{hypotenuse}}$ **Cos** $\theta = \frac{\text{adjacent}}{\text{hypotenuse}}$ **Tan** $\theta = \frac{\text{opposite}}{\text{adjacent}}$ **Number of sides** = $\frac{360°}{\text{exterior angle}}$

Extended section

Volume of any shape = cross-sectional area × length (or height)

Sine rule: $\frac{a}{\text{Sin A}} = \frac{b}{\text{Sin B}} = \frac{c}{\text{Sin C}}$ **Cosine rule**: $a^2 = b^2 + c^2 - 2bc \text{ Cos A}$

$$\text{Cos A} = \frac{b^2 + c^2 - a^2}{2bc}$$

Area of triangle = $\frac{1}{2}ab$ Sin C   **Arc length** = $\frac{\theta}{360} \times 2\pi r$ **Sector area** = $\frac{\theta}{360} \times \pi r^2$

Venn diagram formula: $n(A \cup B) = n(A) + n(B) - n(A \cap B)$

Quadratic formula: $x = \frac{-b \pm \sqrt{b^2 - 4ac}}{2a}$

Equation of linear graph: $y = mx + c$ where m = gradient c = intercept of y-axis

Matrices: $\mathbf{X} = \begin{pmatrix} a & b \\ c & d \end{pmatrix}$ **Determinant** $|\mathbf{X}| = ad - bc$ **Inverse** $\mathbf{X}^{-1} = \frac{1}{|\mathbf{X}|}\begin{pmatrix} d & -b \\ -c & a \end{pmatrix}$

Acceleration/deceleration = $\frac{\text{change in speed}}{\text{time taken}}$ **Distance travelled** = area under speed–time graph

Simple interest, $I = \frac{PTR}{100}$ where P is the principal amount, T the time and R the rate.

Grade descriptions

A **grade A** candidate should be able to do the following.

- Express any number to 1, 2 or 3 significant figures. Relate a percentage change to a multiplying factor and vice versa, e.g. multiplication by 1.03 results in a 3% increase.
- Relate scale factors to situations in both two or three dimensions. Calculate actual lengths, areas and volumes from scale models. Carry out calculations involving the use of right-angled triangles as part of work in three dimensions.
- Add, subtract, multiply and divide algebraic fractions. Manipulate algebraic equations – linear, simultaneous and quadratic. Use positive, negative and fractional indices in both numerical and algebraic work. Write down algebraic formulae and equations from a description of a situation.
- Process data, discriminating between necessary and redundant information. Make quantitative and qualitative deductions from distance–time and speed–time graphs.
- Make clear, concise and accurate mathematical statements, demonstrating ease and confidence in the use of symbolic forms and accuracy in algebraic or arithmetic manipulation.

A **grade C** candidate should be able to do the following.

- Apply the four rules of number to positive and negative integers, and vulgar and decimal fractions. Calculate percentage change. Perform calculations involving serial operations. Use a calculator fluently. Give a reasonable approximation to a calculation involving the four rules. Use and understand the standard form of a number.
- Use area and volume units. Find the volume and surface area of a prism and a cylinder. Use a scale diagram to solve a two-dimensional problem. Calculate the length of a third side of a right-angled triangle. Find an angle in a right-angled triangle, given two sides. Calculate angles in geometrical figures.
- Recognise, and in simple cases formulate, rules for generating a pattern or sequence. Solve simple simultaneous linear equations in two unknowns. Transform simple formulae. Substitute numbers in more difficult formulae and evaluate the remaining term. Use brackets and extract common factors from algebraic expressions.
- Construct a pie chart from simple data. Plot and interpret graphs, including travel graphs, conversion graphs and graphs of linear and simple quadratic functions.

A **grade F** candidate should be able to do the following.

- Perform the four rules on positive integers and decimal fractions (one operation only) using a calculator where necessary. Convert a fraction to a decimal. Calculate a simple percentage. Use metric units of length, mass and capacity. Understand the relationship between mm, cm, m, km, g and kg. Continue a straightforward number sequence.
- Recognise and name simple plane figures and common solid shapes. Find the perimeter and area of a rectangle and other rectilinear shapes. Draw a triangle given three sides. Measure a given angle.
- Substitute numbers in a simple formula and evaluate the remaining terms. Solve simple linear equations in one unknown.
- Extract information from simple timetables. Tabulate numerical data to find the frequency of given scores. Draw a bar chart. Plot given points. Read a travel graph. Calculate the mean of a set of numbers.

Index